Care of the Neurologically
Handicapped Child

WITH CONTRIBUTIONS BY

Suzanne Busch, B.A., M.A.
Educational Consultant, Psychology Laboratory
Department of Pediatrics
Washington University School of Medicine

Mary Pat Hakan, B.S.
Formerly Senior Occupational Therapist
St. Louis Children's Hospital

Ronald M. Lending, B.A., LL.B., J.D.
Formerly General Counsel for Friends of the Retarded

Janet Walker McCann, B.A., M.S., CCC-SP
Speech Pathologist
Washington University School of Medicine

Debra L. Strobach, B.S., M.A.
Intructor in Physical Therapy in Preventive Medicine
Washington University School of Medicine
Senior Physical Therapist
St. Louis Children's Hospital

Valann Tasch, R.N., M.S.N., C.P.N.P.
Clinical Nurse Specialist, Pediatric Neurology
St. Louis Children's Hospital

Carol Weisman, M.S.W., A.C.S.W.
Clinical Social Worker
Children's Hospital, National Medical Center
Washington, D.C.

Care of the Neurologically Handicapped Child

A BOOK FOR PARENTS AND PROFESSIONALS

Arthur L. Prensky, M.D.

The Allen P. and Josephine B. Green
Professor of Pediatric Neurology
Professor of Pediatrics and Neurology
Washington University School of Medicine
St. Louis, Missouri

Helen Stein Palkes, M.A.

Associate Professor of Psychology in Pediatrics
Washington University School of Medicine
Director of Psychology Laboratory
St. Louis Children's Hospital

New York   Oxford
OXFORD UNIVERSITY PRESS
1982

Copyright © 1982 by Oxford University Press, Inc.

Library of Congress Cataloging in Publication Data

Prensky, Arthur L.
  Care of the neurologically handicapped child.

  Bibliography: p.
  Includes index.
  1. Pediatric neurology. 2. Developmentally
disabled children—Care and treatment. 3. Developmentally disabled children—Services for.
I. Palkes, Helen S. II. Title. [DNLM: 1. Nervous
system diseases—In infancy and childhood.
2. Handicapped. WS 340 C271]
RJ486.P73   362.1'968'088054   81-2186
ISBN 0-19-502917-8   AACR2

Printing (last digit): 9 8 7 6 5 4 3 2 1
Printed in the United States of America

To *Lawrence A. Palkes and Pearl Prensky*

PREFACE

It was not too long ago that children with chronic neurological problems were either quickly placed in institutions or cared for by their families without community help. These unfortunate children were thought to be beyond help, although a few might be lucky enough to develop sufficient skills to become semi-independent. Times have changed. Knowledge about the causes and treatment of many neurological handicaps has expanded; many children who would have died early in life are saved, and the population of handicapped children in need of new services has grown. The capability to provide services no longer resides only with the physician; psychologists, teachers, therapists, nurses, social workers, and lawyers now provide help for this population. The proliferation and fragmentation of services is not only confusing to patients and parents, it sometimes confuses professionals whose areas of expertise may overlap or who may know relatively little about the new skills developed in other specialties.

This is a book for parents, but it is also for professionals. We have discussed the types of training given to members of each of the major disciplines concerned with serving the neurologically handicapped child, why a child would be placed under their care, and what services these professionals can be expected to supply. The first two chapters discuss normal and abnormal development and how an abnormality may come to the attention of the child's parents or the medical or educational community. The next chapters outline the services provided by various professionals. The book concludes with a group of chapters that discuss services provided for the seven most common neurological handicaps: epilepsy, cerebral palsy, birth defects, mental retardation, neuromuscular disease, learning and language disabilities, and hyperactivity.

We have tried to provide information without being overly technical. This book is not meant to be used as a handbook for parents or as a teaching manual. On the other hand, some specific information is needed if parents are to ask appropriate questions so that they can make intelligent decisions. A family needs knowledge if it is to develop the attitudes and skills necessary to build an environment in which the handicapped child will thrive. Knowing what the handicapped child is capable of and what his potential capabilities are is the best assurance that parents will develop realistic and constructive attitudes about the child's future. Such knowledge is often difficult to acquire, and the process can be painfully long. Moreover, gaining knowledge requires help from specialists: physicians, psychologists, teachers, therapists, social workers, and nurses. Unfortunately, not all of these specialists may agree with one another, and many of them do not communicate with parents as fully as they should. Our purpose in writing this book is to inform parents and professionals; to identify the different specialists both are likely to come into contact with; and to describe the duties, responsibilities, and services parents can reasonably expect to obtain from each of the professionals who will help them care for their child.

This book is written with a distinct point of view. We feel that parents cannot be passive onlookers at any stage in the care of their child. They should be involved in the evaluation of the child's problems and in the planning of treatment. Caring for a handicapped child is hard work. This book is not meant to be a series of shortcuts for distraught parents; it is intended to help them ask the right questions and to define where their energies and efforts may be best applied to help their children.

Providing a handicapped child with the best possible care requires more than the attentions and expertise of various medical professionals. It requires responsible parents who are interested in their child's welfare; who are willing to seek further information about a medical opinion they disagree with but are also strong enough to stop searching for a more favorable opinion and accept confirmation of a tragic and unwelcome diagnosis; who, given time and proper advice, are capable of developing reasonable goals for their child; who will join in the fight for funds, so that their child and others will be assured an appropriate education, necessary therapy, or needed appliances; and yet who can be objective, who can look at therapies and at educational, vocational, and recreational programs with a critical eye, judging the merits of such programs on the basis of hard evidence, rather than on unsupported, anecdotal claims of "marvelous gains" made by other children. Most of all, a handicapped child needs a family that is able to be warm and loving, yet not overindulgent or overprotective. Parents who are able to be firm and who do not protect the handicapped child from *all* experiences that may be painful or difficult encourage optimal physical and social development. We also feel that the handicapped child must take his place as part of the family unit and that his parents have an obligation to protect the well-being of other family members. Excessive attention devoted to the handicapped child at the expense of his brothers and sisters may only be of slight benefit to him (or even psychologically detrimental); it may seriously impair the emotional adjustment of the siblings. In the end this could only be harmful to every member of the family. Finally, good care requires parents who are willing to learn how to cope with the extra demands on their finances, time, and energy, but who also recognize that if their marital relationship is to endure the extra stresses, they will have to set aside private time for each other, for relaxation, for recreation, a time to talk and to love, so that the marriage is protected and the handicapped child does not become the pivotal and perhaps potentially destructive force in the marital relationship. Loss of private time between the marriage partners often leads to inappropriate hovering over the handicapped child, martyrdom on the part of the mother, and isolation and feelings of rejection on the part of the father. Such a situation almost invariably results in a poorly behaved, dependent, demanding child who is unable to make appropriate social adjustments outside the home.

The professional, in turn, must be attuned to the needs of the entire family and not just the child. To this end we hold that the family physician is the professional who is usually in the best position to advise *all* members of the family about the needs of a handicapped child in relation to their own needs. Whenever possible he should be the central professional figure responsible for the child's care. If the family physician prefers not to assume this responsibility then it must fall upon some other specialist, but each family should look to *one* professional who is a dominant figure in their child's care and to whom they can talk about the family's needs as well as those of the child.

We hope this book helps to inform parents and stimulate them to seek out professionals who will answer their questions. We also hope that it helps inform those of us who work with neurologically handicapped children to better understand what members of other disciplines can contribute.

We wish to thank Mrs. Joan Trunnell, Mrs. Marilyn McLauchlan,

Dr. Philip R. Dodge, and Dr. Mark Stewart for their constructive comments upon reviewing this manuscript. Mrs. Rachel Knipp, Mrs. Dorothy Drezek, and Mrs. Sheila Prensky helped immensely with editorial tasks. Finally, the work could not have been completed without the aid of Mrs. Aileen Derhake, who spent many hours typing the text in its numerous revisions.

St. Louis   A. P.
February 1981   H. P.

CONTENTS

Care of the Neurologically
Handicapped Child

1

NORMAL AND ABNORMAL DEVELOPMENT

The psychological and behavioral development of infants and children is geared to their ability to receive, assimilate, store, recall, and integrate information and to use that information to adapt to the environment. Over time, the baby's reflexes change, new motor skills develop, and his use of the special senses will improve. Ultimately, the normal child develops verbal communication and uses learned information to solve new problems. This increase in the repertoire of responses available to an infant and child stems from growth of the brain and an increase in the number and complexity of connections between nerve cells. These interrelationships are summarized in Table 1.1.

WHAT IS NORMAL?

At birth, the normal baby most often lies with limbs flexed and hands fisted. When hungry, the baby has a good suck and swallows well; his cry is lusty, and when alert but not crying, he will startle to a loud sound and blink to light. By four weeks of age, he may hold his head up momentarily and will usually look at a face. By six weeks, when on his stomach, he will intermittently raise his head to about 45 degrees and smile; he will often follow a moving object with his eyes. Smiling, cooing, and visual following improve by eight weeks, and during this time flexion postures become less prominent and the hands begin to open more freely (Table 1.1). By three months of age, the baby will turn toward sounds and can hold his head and shoulders up when placed on his stomach. His hands are now open much of the time, and when an object is placed in his hand, he may hold it but does not

Table 1.1. Summary of postnatal human development[a]

Age	Visual and motor function	Social and intellectual function	EEG	Average brain weight (g)[b]	Total DNA (mg)[c]	Degree of myelination[d]	Degree of cortical dendritic organization[e]
Birth	Reflex sucking, rooting, swallowing, and Moro reflexes; infantile grasping; blinks to light	—	Asynchronous; low voltage 3-5 Hz; periods of flattening; no clear distinction awake or asleep	350	660	Motor roots +++; sensory roots ++; medial lemniscus ++; superior cerebellar peduncle ++; optic tract ++; optic radiation ±	Frontal gyrus; middle layer III; intersections (branch points) 175 (100)
6 weeks	Extends and turns neck when prone; regards mother's face, follows objects	Smiles when played with	Similar to birth records with slightly higher voltages; rare 14 Hz parietal spindles in sleep	410	800	Optic tract ++; optic radiation +; middle cerebral peduncle ±; pyramidal tract ±	—
3 months	Infantile grasp and suck modified by volition; keeps head above horizontal for long periods; turns to objects presented in visual field; may respond to sound	Watches own hands	When awake, asynchronous 3-4 Hz, some 5-6 Hz; low voltages continue; sleep better organized and more synchronous; more spindles but still often asynchronous	515	860	Sensory roots +++; optic tract and radiations +++; pyramidal tract ++; cingulum +; frontopontine tract +; middle cerebellar peduncle +; corpus callosum ±; reticular formation ±	—
6 months	Grasps objects with both hands, will place weight on forearms or hands when prone; rolls supine to prone; supports almost all weight on legs for very brief periods; sits briefly	Laughs aloud and shows pleasure; primitive articulated sounds, "ga-goo"; smiles at self in mirror	More synchronous, 5-7 Hz activity frequent; many lower voltages, slower frequencies; drowsy bursts can be seen; humps may first be seen in sleep	660	900	Medial lemniscus +++; superior cerebellar peduncle +++; middle cerebellar peduncle ++; pyramidal tract ++; corpus callosum +; reticular formation +; associational areas ±; acoustic radiation +	5250 (925)

Age						
9 months	Sits well and pulls self to sitting position; thumb-forefinger grasp; crawls	Waves bye-bye, plays patty cake, uses "dada," "baba"; imitates sounds	Mild asynchrony; predominant frequencies 5-7 Hz and 2-6 Hz, especially anteriorly; drowsy burst frequent; humps and spindles seen frequently in sleep	750	~900	Cingulum +++; fornix ++; others as previously given —
12 months	Able to release objects; cruises and walks with one hand held; plantar reflex flexor in 50% of children	2-4 words with meaning; understands several proper nouns; may kiss on request	5-7 Hz in all areas; usually synchronous; some anterior 20-25 Hz; some 3-6 Hz; humps often seen in sleep and usually synchronous	925	970	Medial lemniscus +++; pyramidal tracts +++; frontopontine tract, +++; corpus callosum +; intracortical neuropil ±; association areas ±; acoustic radiation ++ —
24 months	Walks up and down stairs (2 feet a step); bends over and picks up objects without falling; turns knob; can partially dress self; plantar reflex flexor in 100%	2-3 word sentences, uses "I," "me," and "you" correctly; plays simple games; points to 4-5 body parts; obeys simple commands	6-8 Hz activity predominates posteriorly with some 4-6 Hz seen especially anteriorly; humps in sleep always synchronous	1065	—	Acoustic radiation +++; corpus callosum ++; association areas +; nonspecific thalamic radiation ++ — 9750 (1225)
36 months	Goes up stairs (1 foot a step); pedals tricycle; dresses and undresses fully except for shoelaces, belt, and buttons; visual acuity 20/20/OU	Numerous questions; knows nursery rhymes, copies circle; plays with others	When awake, synchronous 6-9 Hz predominates posteriorly; less 4-6 Hz activity seen; in sleep, spindles usually synchronous	1140	—	Middle cerebellar peduncle +++ —

Table 1.1. (*Continued*).

Age	Visual and motor function	Social and intellectual function	EEG	Average brain weight (g)[b]	Total DNA (mg)[c]	Degree of myelination[d]	Degree of cortical dendritic organization[e]
5 years	Skips; ties shoelaces; copies triangle	Repeats 4 digits; names 4 colors; gives age correctly	When awake, some 9-10 Hz posteriorly; mostly 7-8 Hz with occasional 4-6 Hz; synchronous; drowsy bursts less frequent and often limited to frontoparietal area	1240	—	Nonspecific thalamic radiation +++; reticular formation ++; corpus callosum +++; intracortical neuropil and association areas ++	—
Adult	—	—	When awake, synchronous 9-12 Hz posterior frequencies; rare 7-8 Hz waves; 18-25 Hz waves and low voltage fast anteriorly; with drowsiness, flattening and low voltage theta; spindles and humps in sleep as before	1400	~1500	Intracortical neuropil and association areas ++ to +++	28,650 (4050)

[a] From Dodge, P. R., Prensky, A. L., and Feigin, R. D.: *Nutrition and the Developing Nervous System*, St. Louis, 1975, C. V. Mosby Co.

[b] From Coppoletta and Wolbach, 1933:

[c] From Winick, 1968; adult value estimated from Dobbing and Sands, 1973.

[d] From Yakovlev and LeCours, 1967. Estimates are made from their graphic data ($\pm$ = minimal amounts; + = mild; ++ = moderate; +++ = heavy).

[e] From Schade, Van Backer, and Colon, 1964. Refers to layer III of middle frontal gyrus. Dendritic intersections and branching points (figures in parentheses) are calculated by multiplying ordinate and abscissa. The figures are arbitrary values but give an accurate order of magnitude.

use it in any way. By four months of age, if not irritable, he will shake a rattle that is placed in his hand, bring both hands together, laugh, actively look around, and become excited when shown food or toys. At six months of age he can generally sit unsupported briefly, roll front to back, try to reach for a toy, recognize his parents in contrast to strangers, and frequently will be able to hold the bottle. By eight months, he can sit alone. By ten months, he can often stand and walk clinging to furniture, and by fourteen months, most infants can walk alone. Many infants say two to three words at one year of age and understand simple phrases. They will transfer toys from one hand to another and reach for the toys they want. The infant can also participate in simple games such as peek-a-boo or patty-cake and can stack two blocks. The Denver Developmental Scale (Fig. 1.1), summarizing normal development over the first six years of life, gives normal milestones.

Development is a continuous process associated with a constantly growing and maturing brain. At birth, the brain weighs about 25 percent of what it will weigh during adult life. Brain weight almost doubles by the end of the first year and triples by the end of the second year of life, at which time it has reached about 75 percent of its adult weight. This remarkable growth is not primarily due to an increase in the number of nerve cells; almost all of the nerve cells in the brain are present in their appropriate positions at the time of birth. After birth, nerve cells continue to grow and send out extensions that connect with other nerve cells in an increasingly complicated way (Fig. 1.2). Much of the increase in the weight of the brain is due to the growth of these extensions. Apparently the ever-increasing complexity of connections between nerve cells allows the infant to develop more complicated skills. It is possible that the development of connections between nerve cells is influenced by the sensory stimulation the infant receives.

A second prominent feature of brain growth during infancy and early childhood is the development of a covering around axons, those long extensions of nerve cells that carry impulses away from the nerve cell body to stimulate other cells. This covering contains an unusually large amount of fat compared with other membranes of the brain and is called the myelin sheath. It allows impulses to be conducted much more rapidly to and from the brain, thus facilitating quick communication between sense organs, brain, and muscles. Unlike nerve cells, the cells that make myelin do multiply rapidly throughout the brain after birth. Thus the rate of formation of this particular membrane is especially susceptible to damage in early life.

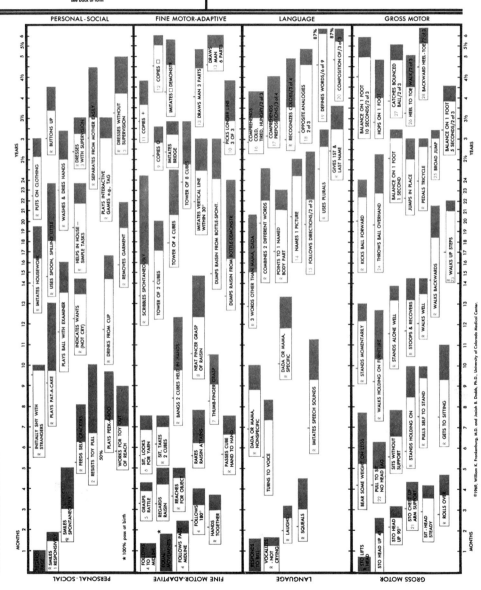

Figure 1.1.

1. Try to get child to smile by smiling, talking or waving to him. Do not touch him.
2. When child is playing with toy, pull it away from him. Pass if he resists.
3. Child does not have to be able to tie shoes or button in the back.
4. Move yarn slowly in an arc from one side to the other, about 6" above child's face.
   Pass if eyes follow 90° to midline. (Past midline; 180°)
5. Pass if child grasps rattle when it is touched to the backs or tips of fingers.
6. Pass if child continues to look where yarn disappeared or tries to see where it went. Yarn
   should be dropped quickly from sight from tester's hand without arm movement.
7. Pass if child picks up raisin with any part of thumb and a finger.
8. Pass if child picks up raisin with the ends of thumb and index finger using an over hand
   approach.

◯	‖	✛	▢

9. Pass any en-
   closed form.
   Fail continuous
   round motions.

10. Which line is longer?
    (Not bigger.) Turn
    paper upside down and
    repeat. (3/3 or 5/6)

11. Pass any
    crossing
    lines.

12. Have child copy
    first. If failed,
    demonstrate

When giving items 9, 11 and 12, do not name the forms. Do not demonstrate 9 and 11.

13. When scoring, each pair (2 arms, 2 legs, etc.) counts as one part.
14. Point to picture and have child name it. (No credit is given for sounds only.)

15. Tell child to: Give block to Mommie; put block on table; put block on floor. Pass 2 of 3.
    (Do not help child by pointing, moving head or eyes.)
16. Ask child: What do you do when you are cold? ..hungry? ..tired? Pass 2 of 3.
17. Tell child to: Put block on table; under table; in front of chair, behind chair.
    Pass 3 of 4. (Do not help child by pointing, moving head or eyes.)
18. Ask child: If fire is hot, ice is ?; Mother is a woman, Dad is a ?; a horse is big, a
    mouse is ?. Pass 2 of 3.
19. Ask child: What is a ball? ..lake? ..desk? ..house? ..banana? ..curtain? ..ceiling?
    ..hedge? ..pavement? Pass if defined in terms of use, shape, what it is made of or general
    category (such as banana is fruit, not just yellow). Pass 6 of 9.
20. Ask child: What is a spoon made of? ..a shoe made of? ..a door made of? (No other objects
    may be substituted.) Pass 3 of 3.
21. When placed on stomach, child lifts chest off table with support of forearms and/or hands.
22. When child is on back, grasp his hands and pull him to sitting. Pass if head does not hang back.
23. Child may use wall or rail only, not person. May not crawl.
24. Child must throw ball overhand 3 feet to within arm's reach of tester.
25. Child must perform standing broad jump over width of test sheet. (8-1/2 inches)
26. Tell child to walk forward, ⚬⚬⚬⚬⚬➔ heel within 1 inch of toe.
    Tester may demonstrate. Child must walk 4 consecutive steps, 2 out of 3 trials.
27. Bounce ball to child who should stand 3 feet away from tester. Child must catch ball with
    hands, not arms, 2 out of 3 trials.
28. Tell child to walk backward, ◄⚬⚬⚬⚬⚬ toe within 1 inch of heel.
    Tester may demonstrate. Child must walk 4 consecutive steps, 2 out of 3 trials.

DATE AND BEHAVIORAL OBSERVATIONS (how child feels at time of test, relation to tester, attention
span, verbal behavior, self-confidence, etc,):

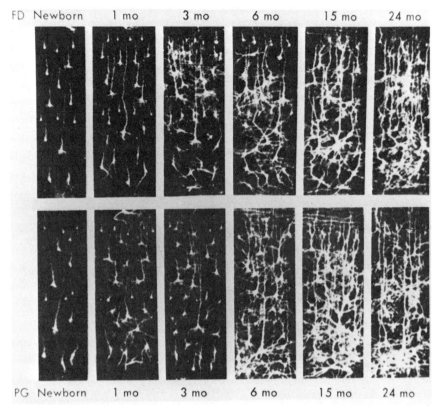

Figure 1.2. Growth of dendritic and axonal plexus in supralimbric homotypical sample areas FD (frontal granular) and PG (parietal granular) from birth to age 2 years. (From Conel, J.L.: The postnatal development of the human cerebral cortex, vols. I–VI, Cambridge, Mass., 1939–1959, Harvard University Press. Reprinted by permission.)

Admittedly, our summary of normal development is an oversimplification. However, it presents many of the important features of development that a parent must understand to comprehend the problems of a disabled child:

1. Biological, intellectual, and behavioral development is an ongoing process that continues throughout life, although at a reduced rate in later years. The biological development of the nervous system has its beginnings in the first weeks of intrauterine life but is far from complete at birth. Most, if not all, of the easily tested and observed reactions of a newborn infant are reflex and require no thought on the part of the baby. In fact, they can be seen in infants who totally lack

the brain regions that are used for "thinking." Even in the first months of life, everything that we view as representing normal development relates predominantly to control of the infant's muscles and development of manual skills. Poor motor control is not necessarily predictive of later defects in intelligence, although they are often associated. Similarly, the achievement of normal motor milestones does not necessarily mean that a child will be intellectually normal or that he might not have severe problems with learning or behavior later in childhood.

CASE 1.1

LS is a fifteen-month-old female. Her parents are concerned that her development may be delayed because she did not begin to roll over until she was eight months old; she did not sit alone until she was eleven months old; and now at fifteen months she is just beginning to pull up to a stand and to walk holding onto furniture. An older sister was able to sit at five months and walked alone at eleven months. The discrepancies between the development of the two girls worried the parents and led them to seek professional advice. Upon questioning, it was found that the child began reaching for objects when she was about five or six months of age and was transferring them from one hand to the other at eight months of age. She now can hold a cup with both hands and stack three blocks; she plays games such as patty-cake and peek-a-boo. She understands a number of single words and has a four-word vocabulary.

Comment:

Do the parents of LS have a cause for concern? They certainly do. Although LS's sister was slightly precocious in both sitting and walking, there is no doubt that the ability of LS to control the movements of her trunk and lower limbs is delayed when compared with the rate of development of the "average" child. Does this mean that LS is retarded? LS's progress has been retarded with regard to the development of specific motor skills, particularly sitting and standing. There is no reason to suspect at this time that LS is mentally retarded. The word "retarded" is often used as an adjective to describe a delay in development of one or more specific functions. When used in this sense, it should not be confused with mental or psychomotor retardation, in which there is a delay in the acquisition of cognitive and social skills. Language and socialization appear to be progressing normally. In fact, LS uses her upper extremities reasonably well for a child of her age.

Upon examination, LS was found to be perfectly normal except for her lower extremities. Her legs were stiff and difficult for the doctor to move. Her arm reflexes were normal but the reflexes in her legs were very brisk. The history and examination suggest that LS has a form of cerebral palsy known as spastic diplegia in which there is damage to limited areas of brain that are particularly concerned with the motor control of the legs and of the trunk. LS could be given a number of disparaging labels that might in themselves affect her future development. The term "cerebral palsy" often evokes unpleasant images unless we remember that it is used to describe a motor deficit that results from a static (nonprogressive) lesion of the brain and that it is not meant to describe the child's mental abilities. LS could also be called "brain-damaged." There is no question that a part of LS's brain has been injured. But a brain-damaged infant or child may be able to lead a normal life if the area of damage is restricted. LS certainly has a problem. She and her family will require professional assistance to help her learn to use her lower extremities to their fullest capabilities in the future. However, it is extremely important that LS's parents be assured that she is developing normally in other ways. She should be treated as a healthy infant and receive the same attention, stimulation, and discipline as her normal sister.

---

2. The nervous system of all but the severely damaged infant continues to develop to some degree after birth. As a result, it is to be expected that virtually all infants with neurological damage will learn some skills as they grow older. Unfortunately, the rate at which they learn these skills may be so slow that they fall further and further behind a normal child of the same age. If a physician evaluates an infant only once, it is often difficult for him to predict the rate at which the baby will develop. Furthermore, the rate of development may vary. There are rare children who develop at a slightly slower rate than normal for the first six months of life only to enter a phase of more rapid progress toward the end of the first year.

3. Infants who suffer intrauterine or perinatal injury or insult that results in brain damage at or before birth, or even shortly after birth, may not develop skills at a normal rate. However, once skills are established, they are generally not lost unless there is a continuing insult to the nervous system. This is extremely important diagnostically, because the loss of acquired skills (degeneration or deterioration) is usually associated with inherited diseases. This has implications not only for the affected child but for other members of the family.

4. Measures of brain growth and charts of motor, social, and verbal

development often fail to consider the great variability that exists in normal development even among individuals in the same family. The Denver Developmental Screening Test (Fig. 1.1), which is widely used by pediatricians and other physicians to evaluate development in infancy and early childhood, attempts to define a time interval during which an ability should develop. When testing the ability of a child to visually follow an arc of 180 degrees, for example, we find that infants normally develop this skill between six and seventeen weeks. Similarly, the ability to sit without support may be acquired before five months or as late as eight months of age and still be considered within the normal range. Twenty-five percent of children can combine two different words by fourteen months of age, but it is only at twenty-four months of age that 90 percent of children can successfully complete this test. Thus the acquisition of this skill could vary from one child to another by as much as ten months and still fall within the "normal" range.

WHAT IS ABNORMAL?

Since there are such wide variations in attaining normal milestones, how do concerned parents decide when to seek professional help because they fear their child is not developing normally? First, let us point out that extreme deviations from normal will probably be noted by the baby's physician—if not in the first few months of life, certainly soon thereafter. Second, children who are globally retarded will usually be delayed in reaching all of their developmental milestones, not just one or two. This will be more obvious if their retardation is moderate or severe. A medical evaluation is apt to provide definite information when an infant is behind in reaching all of his milestones, is losing previously acquired functions, or is very delayed in a specific motor or behavioral act, e.g., not using one side of his body well while using the other normally or developing normal motor and visual skills but not speaking by twenty-four to thirty months of age.

Certain aspects of slow development are apt to be noted by parents first and should then be brought to the attention of a physician. For example, if a mother notes that her child's motor development is normal at three to four months of age, but he seems not to attend visually, it suggests that there is a specific defect in the visual system, and the doctor should be informed immediately. In this case, it is urgent that the child be brought to his physician; if the problem involves the eye, as is likely, it may be treatable. Experience has shown that children adjust better to most visual problems if the defect is recognized and treatment begun early in life.

The same may be said of hearing. If a child responds to visual stimuli but does not respond to sounds and fails to develop speech at a normal age, a physician should be informed. It is quite possible that there is a defect in hearing that could be corrected by the use of a hearing aid. Early language training can allow a child to understand the language and even in some instances to learn to speak well, even if his hearing is defective or absent. Any parent who is concerned that his child has an abnormality in hearing or vision should insist that the child be evaluated by a specialist.

Certain infants and children exhibit obvious symptoms that are immediately recognized by parents as abnormal and create great concern. The most dramatic of these are convulsions, and in this case, parental concern is quite appropriate. The safety of the child during the seizure must be considered, as well as the possibility that the seizure is a symptom of a serious underlying disease. However, after the convulsions have been controlled and the question of a serious disease of the brain has been excluded, parental concern is sometimes excessive. Seizures arise from an area of abnormal brain tissue; this can be a minute area of the brain although the seizure may be dramatic and involve all parts of the body. Thus, the presence of seizures does not always mean that a child will develop abnormally. Most physicians, finding nothing else abnormal about the child, will reassure the parents and recommend that their child be allowed, with few exceptions, the freedom and responsibilities given other children their age.

CASE 1.2

At age fifteen months, DF developed an ear infection with a temperature of 104°F. Shortly after the onset of the illness, he had a three-minute seizure with jerking of all four extremities. When he was brought to the emergency room of the hospital, he was alert, and his examination was entirely normal except for evidence of an ear infection. He had developed normally until this episode. His father, two paternal uncles, and a paternal cousin had had seizures with fever when they were young children. One week after the convulsion, DF had a brain wave (EEG) that was considered mildly slow for his age, but there was no evidence of focal or paroxysmal activity. Because of the abnormal EEG, DF was said to be suffering from epilepsy and was treated with phenobarbital. His parents were concerned about the diagnosis and sought a second opinion.

Comment:

At the time the diagnosis was made, there was no reason to state that DF was suffering from epilepsy. He had had only a single seizure and that had been accompanied by a fever. The appropriate diagnosis was a febrile seizure, not epilepsy. The correct diagnosis was made by the consulting physician. The fact that DF's EEG was slow one week after his attack was meaningless. In many laboratories, up to 15 percent of normal children are found to have a "slow" EEG. Therefore, a "slow brain wave" does not mean DF has epilepsy.

It is appropriate to ask whether having had one febrile seizure, DF will be more likely to develop epilepsy than children who have never had a seizure. DF's parents were also concerned that he might be retarded in his mental development. They were told that his chances of developing epilepsy were slightly higher than those of a child who has never had a convulsion with fever, but that they were very slim. On the other hand, considering his age and his family history, his chances of having another convulsion with further episodes of high fever were greater than one out of five. Thus, treatment with phenobarbital was felt to be appropriate by the consultant. The family was assured that DF's mental development was normal and that he had no greater chance of developing serious school problems than any other neurologically normal child of his age.

Physicians generally have little difficulty recognizing the child whose development deviates from normal in one or more spheres of activity. However, they often have great difficulty explaining the meaning of such deviations to parents. Understandably, parents want to know whether a child with seizures or motor problems will have normal intelligence. Because different facets of the developing behavior of a child, such as movement and articulation, may not relate to one another or to the ability to learn later in life, the physician often must hedge his responses or postpone answering a parent's questions. What he should *not* do is pass over a deviation from normal as "a lag" and either not mention the abnormality to the family or tell them "he'll grow out of it." Parents have a right to know about any observations made about their child, and they should be given some rough idea whether a developmental abnormality will be meaningful. When that is not possible, they should be told that they must be patient, and they should be given some idea as to when more specific predictions can be made.

REFERENCES

Brown, S. B., and P. K. Sher. 1975. The neurologic examination in children. *In* K. F. Swaiman and F. Wright, eds. The practice of pediatric neurology. C.V. Mosby Co., St. Louis, pp. 9–50.

Dodge, P. R. 1975. Neurologic history and examination. *In* T. W. Farmer, ed. Pediatric neurology. Harper & Row, Hagerstown, Md., pp. 1–43.

Dodge, P. R., A. L. Prensky and R. D. Feigin. 1975. Normal cerebral maturation. Part I *in* Nutrition and the developing nervous system. C.V. Mosby Co., St. Louis, pp. 3–160.

Frankenburg, W. K., and J. B. Dodds. 1967. The Denver Developmental Screening Test. J. of Pediatrics, 71: 181–191.

Illingworth, R. S. 1980. The development of the infant and young child: Normal and abnormal, 7th ed. Churchill, Livingstone, Edinburgh and New York, p. 320.

Kagan, J., R. B. Kearsley, and P. R. Zelazo. 1978. Infancy: Its place in human development. Harvard University Press, Cambridge, Mass., p. 462.

Knobloch, H., and B. Pasamanick, eds. 1974. Gesell and Amatruda's developmental diagnosis: The evaluation and management of normal and abnormal neuropsychologic development in infancy and early childhood, 3rd ed. Harper & Row, Hagerstown, Md., p. 538.

Lund, R. D. 1978. Development and plasticity of the brain. Oxford University Press, New York, p. 370.

Paoletti, R., and A. N. Davison. 1971. Chemistry and brain development. Plenum Press, New York, p. 457.

Paine, R. S., and T. E. Oppe. 1971. Neurological examination of children. Spastics International Medical Publications, Lavenham Press, Lavenham, England, p. 279.

Piaget, J. 1952. The origins of intelligence in children. W. W. Norton and Co., New York, p. 419.

Purpura, D. P., and G. P. Reaser. 1974. Methodological approaches to the study of brain maturation and its abnormalities. University Park Press, Baltimore, p. 155.

Robertiello, R. C. 1975. Hold them very close, then let them go: How to be an authentic parent. The Dial Press, New York, p. 176.

Touwen, B. 1976. Neurological development in infancy. William Heineman, Ltd., London, p. 150.

2

THE DISCOVERY
OF A NEUROLOGICAL
HANDICAP

Webster defines the term "handicap" as "a disadvantage that makes achievement unusually difficult." In this book, we will be talking about neurological handicaps, which we shall define as "any condition, congenital or acquired, that alters the normal functioning of the nervous system during the patient's lifetime and thus hinders achievement." Both of these definitions suggest that achievement is possible, *and it is*. It would be a severe handicap indeed that would totally abolish any chance of acquiring new skills. On the other hand, skills that might be acquired effortlessly by the average child may be achieved only after great work by a handicapped boy or girl, and some skills may be out of their reach. Other goals may have to be set and other skills substituted.

The handicaps we shall be discussing in this book are epilepsy, mental retardation, cerebral palsy, neuromuscular disorders, special learning disabilities, hyperactivity, sensory deficits, language disorders, and birth defects.

HANDICAPS EVIDENT AT BIRTH

There are three ways in which a neurological handicap may make itself evident at the time of birth. First, there may be obvious signs of damage to the nervous system that will impress both the family and the physician. Seizures or problems with movement, such as paralysis, or difficulty breathing are among the most striking of these signs. There may or may not be an associated history of a complicated pregnancy or delivery. Second, a child may have physical features, such as abnormalities in the development of the face, the ears, or the hands, that may be associated with damage to the nervous system. The child may have the distinctive features of mongolism (trisomy 21), for example, or he may have an unusually large head or a cyst protruding from a defect in the spinal column. Third, a child may appear perfectly normal but be identified by routine screening as suffering from a metabolic disorder, such as phenylketonuria, that could result in mental retardation. In metabolic disorders, the infant or child is not able to make normal use of certain chemicals that he needs for growth. These are substances that he either makes within his body or extracts directly from food. The inability to properly metabolize one or more chemicals in the body is often inherited. During pregnancy, the mother's body may perform this task for the infant; thus, the baby often appears entirely normal at birth. Only after he must build up or break down chemical compounds in his body when separated from his mother do symptoms occur.

Physicians are often reluctant to discuss with the family the implications of a neurological handicap that is suspected or is present at the time of birth unless they can diagnose a specific disease or syndrome. Without a specific diagnosis it is difficult to predict the child's future. Children with severe respiratory problems or seizures at birth will often develop normally, while children with relatively minor problems may at a later date show marked delay in their development. If there is no definite disease, a physician can only base his prediction about a child on statistical probabilities, that is, upon information about development gathered from a large group of patients with similar histories or physical findings. Statistics, however, cannot predict the future of any given child with certainty.

Complications seen at birth may be the result of a variety of causes. Viral infections, particularly early in pregnancy, can result in psychomotor retardation, cerebral palsy, blindness, and deafness. However, even when such an infection can be proved, only a small percentage of children will develop these neurological handicaps. The most serious of these viral infections is rubella (German measles). About one-third of the infants born to mothers who had German measles in the first trimester of pregnancy have severe neurological damage. If a pregnant mother suffers from malnutrition, hormonal disorders, is exposed to poisons such as mercury or other toxic metals, or takes prescribed drugs such as some anticonvulsants, her unborn child (fetus) is also more susceptible to injury. Recently, excessive use of alcohol and cigarette smoking during pregnancy have been implicated in producing children with brain damage. Severe hypertension during pregnancy also subjects the fetus to extra risk. If the placenta is misplaced or poorly attached to the wall of the uterus, the baby can be deprived of necessary nutrition from its mother. Vaginal bleeding is often seen during the pregnancy or at the time of labor as a symptom of these problems.

Infants who receive too little oxygen during the birth process can be born blue (cyanotic). Such a child may have to be given respiratory support at birth. It is common to classify children at birth according to the Apgar score, which indicates an infant's physical condition at birth, at one and five minutes after delivery. Respiration, color, heart rate, muscle tone, and response to stimulation are each rated on a scale of 0 to 2. A perfect Apgar score is 10. Very low Apgar scores (in the range of 0–3) are associated with a greater risk for neurological handicaps such as retardation and cerebral palsy. Nevertheless, many distressed infants with Apgar scores in this range will develop normally.

CASE 2.1

RT was born at term with a birth weight of 2800 gm (6 lbs 3 oz). Upon delivery, the umbilical cord was found wrapped around the child's neck; she had difficulty breathing, and required artificial respiration for approximately five minutes after birth. Her Apgar score at one minute of life was 2; at five minutes it was 5. She was placed in an incubator for the first four days of life and required careful observation because she stopped breathing briefly on several occasions. She had no other problems. Her examination and EEG were normal at three weeks of age. Her parents were quite concerned that she would be retarded, and they were upset that they could get no clear-cut predictions about her future from their doctor.

Comment:

In this instance, the parents are wise to be concerned, and their child's physician is even wiser to give them no definite predictions, although both are probably unhappy with each other. A child with this kind of history has a slightly greater risk of developing a neurological handicap later in life than a baby born of a perfectly normal pregnancy and delivery. On the other hand, since RT had no evidence of severe neurological damage, there is no way of predicting whether she will develop normally. At this time, patience on the part of the parents and careful observation on the part of the doctor is indicated. The ability to define and list the insults or stresses to which an infant was subjected at or before birth does not constitute a diagnosis of a disease; it is the recognition of a transient disorder that may or may not result in a disease of the nervous system.

Extreme prematurity with birthweights under 1500 grams (3 lbs) or a low birth weight for gestational age (under 2500 gm [5 lbs] for a forty-week or full-term infant) are also associated with a higher risk for development of neurological handicaps. Infants who at birth have poor muscle tone (floppy babies) and are less active than normal newborns, or infants who in response to slight stimuli show increased movements, such as tremors or convulsions, are at greater risk. However, these conditions need not result in a damaged nervous system.

The chance that an infant will develop poorly increases if he suffers from a combination of insults. For example, a low-birthweight infant who had seizures in the first day of life and was born to a mother who had high blood pressure during pregnancy has a greater chance of doing poorly than an infant with only one such problem. Even under these circumstances, however, we are dealing with probabilities. Only the passage of time allows a physician to predict what will happen to a specific infant unless he can make a definite diagnosis of an underlying disease or permanent injury to the nervous system.

Such a diagnosis can sometimes be made at birth, as in the case of birth defects. The failure of closure of the vertebrae (the bones that surround the spinal cord) and the bulging through this opening of the coverings of the cord and the cord itself, combined with partial or total paralysis of the lower extremities, is evidence of a neurological deficit known as myelomeningocele, a condition that will not disappear even though the sac itself can be removed and the bony defect covered with skin. Similarly, the presence or rapid development after birth of hydrocephalus (an excessively large head at birth that continues to grow rapidly and contains excessive amounts of fluid), often results in permanent damage unless there is surgical intervention. At the other end of the spectrum is the child with an extremely small head (at least three standard deviations below the mean for age); this infant is also apt to develop very slowly.

Evidence of abnormal formation of other parts of the body, such as widely spaced eyes, fused fingers, and low-set, poorly formed ears without the usual ridges and creases, when found in combination, are often better predictors of later developmental handicaps than are neurological signs at birth. Abnormalities in the development of the limbs or of the face generally indicate some insult to the infant early in pregnancy. At times, particular constellations of physical features may suggest that the child has a specific disorder, such as chromosomal abnormality.

CASE 2.2

TB was born at term of a normal pregnancy. At birth, he was noted to have a somewhat flat, round face, upward slanting eyes and a protruding tongue. His ears were simple, his neck was short, and his fifth finger curved in. He had creases that crossed his entire right and left palms. His developmental examination was normal except for decreased muscle tone. Even though there had been no difficulty with the pregnancy or delivery and the child's only neurological abnormality was poor muscle tone, his parents were told that he would probably be retarded. It was suggested that a chromosome study be done because the child had the physical features of mongolism (trisomy of chromosome 21). Indeed, he had an extra chromosome 21.

Comment:

In this patient, the diagnosis of probable psychomotor retardation was made by the observation of physical abnormalities that had no direct relation to the nervous system. These abnormalities suggested a trisomy 21 (the most common of all chromosomal abnormalities, present in one out of 500 to 800 live births). Just as not all normal babies are the same, not all mongoloid children are the same, although virtually all are retarded. These children vary significantly in their degree of retardation. Many can be trained to care for themselves with minimal supervision, and occasionally a child with trisomy 21 can be educated.

Clusters of physical abnormalities that form a clinical syndrome often have a definite pattern of inheritance that families should be made aware of. In some instances when a chromosomal abnormality is discovered in a child, it is essential that the parents' chromosomes also be examined.

It is equally important to remember that most children who have unusual physical features have a normal nervous system. Perhaps the most dramatic of these children have been the thalidomide babies, who have one or more malformed extremity yet a normally functioning brain.

Metabolic screening of newborns is a relatively recent technique, but it is now required in almost all states for the diagnosis of phenylketonuria. In fact, to be sure that the child does not suffer from this disease, it is best that he be screened about six weeks after birth as well as in the first days of infancy. In most instances, screening for phenylketonuria is done by state laboratories from samples of dried blood.

CASE 2.3

RD was born after a normal pregnancy and delivery. There was no history of neurological disease in either his mother's or father's family. State law requires that all newborn babies be checked for phenylketonuria. Blood samples are drawn at three days and again at six weeks and sent to a state laboratory for testing. Accordingly, a sample of RD's blood was tested at the designated times. At three days of age, his blood level of the amino acid phenylalanine was four times that of a normal infant, while at six weeks of age it was thirty-five times the normal. Extremely high blood levels of phenylalanine occur in phenylketonuria since the amino acid cannot be properly metabolized. The result from the state laboratory was confirmed. The child was then started on a diet that was restricted in phenylalanine. He has developed normally. At age eight, his I.Q. was estimated at 110, which approximates that of two older siblings.

Comment:

Phenylketonuria is inherited as a recessive disorder. By this we mean that both parents must carry the abnormality, but they need not have any symptoms or signs. Only when a child gets a double dose of the abnormal genetic material—that is, from both father and mother—is the disorder clinically apparent. As a result, when a disease is inherited as a recessive trait, it is not uncommon to find that there is no family history of a neurological disorder. As is the case with many inherited diseases that involve abnormal metabolism, signs of the disorder frequently are not present at birth. These metabolic disorders are very rare. Phenylketonuria, the most common, occurs only once in every 15,000 live births. Routine screening for other metabolic problems generally has not been done because of cost. However, screening for other treatable diseases is increasing, because treatment early in life can often prevent permanent injury to the brain. Admissions of phenylketonuric patients to state mental hospitals or schools for the retarded have almost ceased since routine neonatal screening has become almost universal.

Many rare biochemical disorders can be identified before birth. This testing would be indicated if there is a family history of such a problem. Many of these diseases cannot as yet be treated. However, if such disorders can be identified early enough in pregnancy, the family, based upon their beliefs, will then have a choice of whether or not to terminate the pregnancy.

HANDICAPS EVIDENT IN THE FIRST YEARS OF LIFE

Neurological handicaps need not be present at birth nor need they be associated with an abnormal history during pregnancy, delivery, or the perinatal period. Convulsions, for example, can begin at any time in life and are particularly common in the first six months after birth, decreasing in frequency thereafter until puberty. The earlier in life convulsions begin and the more difficult they are to control, the more likely that patients with this problem will have to be treated throughout their lives and will have associated neurological handicaps. Other abnormalities particularly noticeable in the first years of life are the result of failure to develop normally; initially this is manifested by failure to reach proper motor milestones such as sitting, reaching for objects, or walking. Such a delay is often the first sign of psychomotor retardation. However, it may be an indication of cerebral palsy or a manifestation of a disorder of nerve or muscle, and the child's intelligence may be normal. Early in life it may be difficult to separate intelligence from motor function, and many years may be required to assure a family that their child's developmental handicap is limited to control of movement.

The development of language in the first two years of life is so variable that lack of speech in the presence of normal social development and understanding of language should not provoke great anxiety. However, if it is clear that the child does not understand simple speech or is not interested in sounds, a hearing test is certainly indicated.

Less frequently, an infant or child reaches his milestones at the normal time only to lose these skills gradually as he grows older. This is a much more serious problem. The loss of these skills, if it is not associated with a severe insult to the brain, such as an acute injury, tumor, or meningitis, is almost always associated with inherited diseases in which parts of the nervous system are progressively destroyed.

HANDICAPS EVIDENT AFTER THE FIRST THREE YEARS OF LIFE

With the exception of school failure and seizures, recognition of new neurological handicaps later in childhood is unusual unless the symptoms or signs are part of a recent trauma to the nervous system. School failure may be due to inability to concentrate, a mild intellectual disability that was not previously noticed, or more commonly, a specific learning disability that interferes with a child's ability to learn certain skills only when that individual is confronted by specific tasks.

There are diseases that do occur later in childhood and result in gradual deterioration of the nervous system. Sudden loss of function can occur after trauma or infection just as it can in the first years of life. Convulsions, as noted earlier, can occur at any age and may not signal a serious illness such as a tumor. It is more likely that these seizures constitute a reaction to an injury that occurred many years before. Handicaps (with the exception of epilepsy) that are discovered after the first years of life are less likely to disappear with further maturation than are those that develop early in life.

School problems are usually first apparent to teachers, as might be expected. However, they often become evident as a difficult diagnostic problem because of poor communication between the family and professionals. There are many reasons why a child may not be able to keep up with his classmates. Some of these are emotional or social; others involve neurological handicaps that vary from a poor attention span or a specific learning disability to mental retardation or mental deterioration. If school failures are not identified promptly and an appropriate diagnosis and disposition made, behavioral problems are often superimposed upon intellectual ones as the child tries to compensate for his disability. Many times it requires a highly experienced educator and sometimes an informed physician and psychologist to determine the nature of the disability and plan a therapeutic program. At all times, coordination between specialists and the family is essential. Learning disabilities and their consequences are not restricted to the schoolroom, nor is their treatment.

CASE 2.4

NB was an eight-year-old boy who had always been thought to be normal. His parents were surprised when the boy's third-grade teacher asked them to come in for a special conference. The teacher indicated that she found it impossible to teach NB to read and that when the class was reading aloud, he often became disruptive. Furthermore, when he tried to read aloud and was unable to do so, other children made fun of him and this led to fights. She felt that she could not tolerate the boy in class any longer.

His neurological examination was entirely normal. However, when he was asked to read, he misread letters, particularly "b" and "d," and reversed words, saying "saw" for "was." He had considerable difficulty with early first-grade material. He had similar problems with writing, and he was unable to do simple calculations with the ease expected of a normal second-grade student. However, an intelligence test showed his I.Q. to be within the normal range. The boy's parents were quite upset with his pediatrician because he had always indicated that NB was normal.

Comment:

It was inappropriate (although understandable) for NB's parents to be upset with his pediatrician. There is no way that a specific learning disability can be diagnosed before the basic school skills are introduced. It is rare for a physician to screen for such problems in routine office visits. As a rule, physicians rely on the school or the parents to identify this kind of handicap. Unfortunately, the problem is often ignored at school and the child moved ahead with "social" promotions. NB's parents were justifiably upset to find that their son's learning problem had not been identified before the third grade. Indeed, even at this time it was the boy's behavior problem rather than the teacher's recognition of his learning problem that provoked the special conference. Once behavior deteriorates, it is much more difficult to reestablish the child in a productive educational setting.

WHO DISCOVERS A HANDICAP?

Handicaps that develop at or shortly after the time of birth are generally discovered by the infant's physician. Occasionally, it is the mother who will notice that a child does not suck well or has an unusual cry in the first days of life, but this is rare. Physicians are often reluctant to make any definite statements about the extent of an infant's handicap unless they can make a specific diagnosis. There is, however, every reason to expect a physician to inform the family that a handicap exists and to make provision for the infant to be followed carefully so that the severity of the problem can be assessed and treatment instituted as soon as it is appropriate.

Later in infancy and in early childhood, handicaps may be identified by the family or even by friends or neighbors before the physician is aware of them. This is particularly true of deafness and of visual problems. It is difficult during a brief examination to make a definitive diagnosis of these disorders unless they are obvious. On the other hand, people who live with a child may be aware that there is something wrong with his reactions to verbal or visual stimuli, even though they may not be able to put their finger on the exact problem. Under these circumstances, it is always wise for family members to trust their own instincts, to bring their concerns to the attention of their physician, and if necessary to seek further help.

Repeated convulsions or epilepsy is also a problem in which the physician must depend on a description given him by family, friends, or teacher. It is rare for seizures to occur in the doctor's office. Epilepsy is a diagnosis that is usually made *only* by history. Therefore, patients who have a normal physical examination and electroencephalogram may still have the disorder if the description of the attacks leaves no doubt about the diagnosis. Under these circumstances, for a physician to make an accurate judgment, it is important to have the most detailed information possible—preferably, to have witnesses present during the examination who have actually seen an attack.

REFERENCES

Apgar, V., and J. Beck. 1972. Is my baby all right? Trident Press, New York, p. 492.
Brown, D. L. 1972. Developmental handicaps in babies and young children: A guide for parents. C.C. Thomas, Springfield, Ill., p. 89.
Moore, C. B., and K. G. Morton. 1976. A reader's guide for parents of children with mental, physical or emotional disabilities. U.S. Department of Health, Education and Welfare. DHEW Publication No. (HSA) 77–5290, p. 144.
Weiner, F. 1973. Help for the handicapped child. McGraw-Hill, New York, p. 221.

3

THE FAMILY PHYSICIAN

THE ROLE OF THE FAMILY DOCTOR

We are defining general practitioners, specialists in family practice, and pediatricians as the child's family physician, i.e., the doctor who has primary responsibility for a child's general medical care. These physicians are either the first to notice that a child is not performing normally or the first to be consulted by the family because they have concerns about their child's development. There are many reasons why the parents of a neurologically handicapped child might look upon the family doctor as central to the care of their child. He is the doctor whom they know best, visit most often, and who is nearest to them both geographically and emotionally. Frequently, the family physician cares for other members of the family. Since he is often the first professional who has contact with a child, he is likely to be the first to become aware of problems and discuss them with the family. Even if a specialist is used to treat a child's neurological handicap, the family physician still will be responsible for the child's general medical care. Moreover, if the child lives some distance from a major medical center, the family physician will render any emergency care that may be needed because of the child's neurological disorder.

The professional care of the handicapped child, particularly one with multiple neurological handicaps, can be visualized as a wheel. At the hub is the family physician and from him radiate spokes to other physicians with more specialized training—pediatric neurologists, neurosurgeons, orthopedists, specialists in physical medicine, etc., or other professionals such as psychologists, physical and occupational therapists, social workers, teachers, community health officers, and visiting nurses. There are actually very few neurological problems that require frequent visits to medical specialists or that require a specialist to become the physician primarily responsible for a child's health needs.

There are many reasons why the type of care suggested by the wheel analogy, which most of us consider to be ideal, breaks down and the family physician is replaced as the hub. The child's medical problem may be so rare and complex that it requires the combined efforts of a number of medical specialists who are immediately available to one another for consultation. In such instances, the specialist who is managing the major underlying disorder would assume primary responsibility for the coordination of the care of the child. Other causes lie in the practical needs and the personality of the family doctor or the parents. The needs of a neurologically handicapped child very often cannot be met fully by a periodic examination and a prescription for medication. In fact, except for children who have easily controlled epileptic seizures, it is safe to say their needs can never be met by medication alone. Counseling the family about raising a handicapped child, identifying sources of stress in the family, providing advice about preschool and school programs, and deciding if and when physical and occupational therapy should be initiated are only a few of the physician's other responsibilities. Identifying a child's needs and discussing them with a family is time-consuming. Many doctors are willing to spend this time, but others are not because their fee per visit is small and their practice depends upon caring for large numbers of patients quickly. Some have no desire to be involved in the problems of a chronically ill child and his family and prefer the immediate satisfaction that comes with curing an infection or relieving pain. It is sometimes valuable for parents of a handicapped child to sit down with their family doctor and ask him frankly if he is interested in treating their child's chronic problems.

At the very least, parents have the right to expect that if their family physician is not going to assume the responsibility for the continuing care of their child's neurological problem, he should be willing to cooperate with a group of professionals or a specialist who will assume that role. The family physician is ideally qualified, however, since his knowledge of family interactions will always exceed that of the medical specialists who care for the child.

CASE 3.1

RZ, a 2½-year-old male, was born after a pregnancy complicated by considerable bleeding just before delivery. The baby was blue at birth and breathing poorly; he required artificial respiration for at least thirty minutes. He had seizures in the first week of life and had occasional generalized seizures and many little jerks and head drops thereafter. Nevertheless, by eighteen months he was sitting alone, although he would only occasionally reach for objects and bring them to his mouth. He was unable to transfer objects and showed no reaction to sound, although he seemed to look at objects in the room. His family physician hospitalized him and ordered a number of tests. X-ray studies included computerized tomography (a CT scan), which showed that there was some loss of brain substance, probably due to lack of oxygen at birth. Other studies were normal, with the exception of the brain wave, which showed electrical seizure activity. The child was started on phenobarbital, but his seizures continued, and he also became irritable and drowsy. The family physician then started the child on a second anticonvulsant, also without effect except for increased irritability. The physician was concerned about the child's slow motor and intellectual development and his continuing seizures. He advised the family about community resources that might be of help to them and their child. He arranged for the parents to consult the hospital social worker about financial assistance and about how to obtain help in caring for their child two half-days each week so that they could have some time to themselves. He arranged for the child to enter a therapy program at a regional Cerebral Palsy Center. He also referred the parents to a pediatric neurologist for further advice about the diagnosis and about seizure control.

Upon reviewing the test results and the hospital notes that were sent with the patient, talking to the parents, examining the child, and obtaining an EEG and some chemical studies of his blood, the neurologist felt the child's problems were the result of lack of oxygen at birth. He felt the child's mental function might improve with better seizure control and added sodium valproate to the child's regimen. He saw the child every month until it was clear that the seizures had become infrequent. Thereafter, he suggested the family physician continue to care for the child and wrote him a detailed note suggesting future changes in drugs.

RZ developed slowly but eventually learned to obey single commands and developed a vocabulary of 100 to 150 words. A psychologist tested RZ and found him to be in the trainably retarded range. The psychologist and family physician worked with the local school system to obtain an appropriate classroom placement for the child.

Comment:

In this case, the family physician made use of all the resources in the community, but remained the primary professional caring for RZ. This arrangement benefited the child and the family because the physician clearly understood their needs.

THE FIRST VISIT

We are defining the "first visit" as the one during which the child's doctor becomes aware of a neurological problem. As we indicated previously, it is the family physician who is most likely to be the first to suspect that the child might have a neurological problem. He should then discuss his suspicions with the family and indicate whether he is willing to assume responsibility for the child's continuing care. This initial discussion between family and physician is extremely important, because it often influences how the parents will think about their child's neurological handicap in the future. There is no reason why a physician should not tell a family the truth about their child's condition during an initial discussion. It is extremely important that a family have faith that their doctor has been entirely honest with them. However, this means that in addition to giving parents as much specific information about their child's problems as he can, the doctor must also make clear those areas of the child's illness that he may not understand or can only speculate about. In the case of neurological disorders, these areas are often broad, and this means that a doctor may have to answer "I don't know" or "I have no idea" to many questions. Parents should understand that this does not mean that the doctor is uninformed. On the contrary, it may mean he is a very good physician and that he recognizes the limits of his medical knowledge.

In the discussion with the family following the confirmation of a handicap, there are several questions about the child's condition that most parents will want answered:

1. What is wrong with our child?
2. How did it happen?
3. Will it get worse or better?
4. Are there going to be other problems, such as retardation or seizures?
5. What can we do to help our child? Should we seek further professional advice?
6. If we have any more children, will they be affected?

Certain general principles govern how detailed the answers to these questions can be.

What is wrong with our child? Eventually, a physician usually will be able to describe a child's disabilities to parents even though he may not be able to make a definite diagnosis. By "definite," we mean a diagnosis that establishes that the patient's symptoms and signs are caused by a known disease. It is important to realize how much the practice of medicine still depends upon the use of the Latin language, particularly if a doctor is labeling a group of signs or symptoms. Many "diagnoses" are in reality translations of symptoms or signs into Latin or Greek. For example, a parent says, "I noticed my child doesn't move one side of his body," and the physician states: "He has a hemiparesis." This is a Latin term for paralysis of one side of the body. A child complains of muscle pain and a physician says that he has "myalgia," which describes the complaint in Latin (i.e., a pain in the muscle). A parent tells the doctor that his child has fallen to the ground several times and jerked all over, and the physician then says, "Your child has epilepsy," to which he may append the English adjective "generalized" or the French term "grand mal."

All of the terms used here, which are commonly used in discussions between physicians and patients or their families, such as epilepsy, mental retardation, cerebral palsy, learning disability, or hyperactivity, are accepted methods of describing a symptom or group of symptoms and signs. *They do not refer to a disease.* These symptoms can be caused by many different diseases or insults to the brain, and because they are symptoms or signs and not diseases, they can vary remarkably in their severity. In addition, it is often extremely difficult to predict their course. Parents should ask whether the doctor is putting a label on a symptom or sign or actually describing a disease.

How did it happen? Frequently, a physician will be unable to state how or why a neurological disability came about. At times, the course of the illness or its signs will suggest a definite cause. As noted earlier, these could include a viral infection during pregnancy, a chromosomal abnormality, or possibly an inherited degenerative disease of the nervous system. More often, there will be a history of a suspicious event during pregnancy or delivery, or shortly after birth. However, relating a symptom such as epilepsy or mental retardation to one of these factors is tentative at best. As we indicated in Chapter 2, the association between a single event of this kind and a subsequent handicap in a particular child is uncertain. A large number of handicapped children have not suffered any known insult during pregnancy or delivery, or after birth. Nor do they show any evidence of a genetic or chromosomal problem. Therefore, the questions of "how" and "why" the disorder occurred cannot be answered.

The tendency of families to focus on why or how their child's neurological disability came about and to try to assign cause or blame usually results in no definite conclusions and is of no value to the child. The common denominator of neurological handicaps is brain damage, and the nature and severity of the handicap depends more on which areas of brain and how much of the brain has been injured than it does on the cause of the injury. Undue concern with first causes is one of the most common ways in which a family dissipates time and energy that could be used more productively. Insistent attempts to talk about causes may also alienate the physician, because so much time is thereby wasted in fruitless discussion.

On the other hand, physicians are not immune from this fixation on "how and why," and if a handicapped child's physician is overly concerned with this problem, it may be detrimental to the child's care. Some physicians feel that if they cannot assign a cause to a neurological handicap, they cannot plan a treatment program, and they thus ignore the child's immediate needs. This should never happen. When faced with this attitude, a family should ask to be referred to a specialist for a second opinion.

Will it get better or worse? Prognosis is often a difficult matter for a doctor to discuss during the "first visit." This is particularly true if the child is very young. The younger the infant, the more cautious the doctor should be about predicting outcome. If the damage to the nervous system is extremely severe, or if the child's condition has clearly been getting worse over a period of weeks or months, the doctor can be less guarded about his predictions. The older the child and the longer the handicap has been obvious, the more accurate a family can expect the doctor to be in his predictions.

Are there going to be other problems? It is not easy for a doctor to predict whether a child who has a handicap such as epilepsy or cerebral palsy will also have other deficits unless the child is so severely damaged that it is clear that development will be seriously delayed in all areas. The doctor must usually rely on statistics and suggest that although neurologically impaired children are more susceptible to these complicating problems, there is no way of predicting how a particular child will do. The doctor should help parents accept their child as an individual and define his disability for them as clearly as possible. He should not encourage parents to focus on potential future handicaps. Yet in a sense both parents and physician must walk a tightrope, since children with one neurological handicap are more likely to develop others, and it is important to discover and treat these problems as early as possible. Most of these complications, if they occur, will become apparent to the family or the physician with little effort. Constantly scrutinizing the child for some new problem is apt to be psychologically destructive to the child and to the family.

What can we do to help our child? Love is not enough. Neurological handicaps are not curable in the sense that pneumonia can be cured.

However, they are all treatable to some degree. While treatment often involves medication, in many cases it requires manipulation of the child's environment to encourage him to make increased use of his existing abilities and to stimulate him to learn new skills. Frequently, this kind of treatment should be started early in life. There are very few instances in which a physician should answer this question by saying "Nothing can be done."

If we have other children, will they be affected? Young parents who have a child with a neurological handicap worry about having more children. They are appropriately concerned about the chances that any other children they might have will develop a similar disorder. They are frequently worried about future grandchildren. If the child's symptoms and signs lead to the diagnosis of a definite disease or a specific syndrome, some information can often be given about inheritance. If, as is more common, there is no definite cause for the child's handicap, then only the most general information can be given to a family. For example, if one child in a family has epilepsy, there is a somewhat greater chance that his siblings or his children will also have epilepsy. However, the odds are still approximately fifty to one against either siblings or children of the patient developing seizures unless there is already a strong history of epilepsy in other members of the immediate family. If a child has a myelomeningocele, there is a 5 percent chance (one chance in twenty) that any future child that the mother bears will suffer from the same or a similar defect. If a child is retarded but not deteriorating and no cause for the problem can be found, there is probably only a slightly increased chance that any future children will be retarded. Good statistics are not available, but one can presume that the chances of any future children having a similar problem are under 5 percent. The overall incidence of neurological handicaps in the entire population is between 5 and 10 percent, if one includes language disorders and learning disabilities. These disorders are so common that it would be foolish for a physician to assure a family that none of their future children will have a neurological handicap.

The family doctor should bring up these questions if the family does not. Very often, parents are reluctant to ask questions for fear that the doctor will confirm their worst suspicions. At other times parents do not understand what the doctor has said about their child's condition, but they are too embarrassed to ask him to explain himself in simpler terms. The physician should always be willing to discuss these questions with the family and to give them the information they need to understand their child's condition more clearly.

CONTINUING CARE

The family physician who is willing to undertake the continuing care of neurological handicaps should have certain skills and information. The major asset of the family physician is his continuing relationship with the child and his family. He should know the family well professionally or be willing to take the time to understand the relation between the child's medical problems and the functioning of the family. The physician must understand how to perform and evaluate a neurological examination. He must be aware of the therapies available for the chronically handicapped child, of recent advances in treatment, and of current research related to the child's clinical problem. He should be interested in the child's social and educational development.

A family physician who assumes responsibility for the care of a neurologically handicapped child must be familiar with community resources available for the care of that child. Needed services might include programs in physical and occupational therapy, infant stimulation and early education, and, if necessary, later in the child's life, special education programs in public or private schools. The physician should be able to advise the family and child about vocational rehabilitation and social programs for the handicapped. Many private agencies concerned with specific diseases will help the child and his family by providing some financial aid or medical or social services. It is not reasonable, however, to expect a physician to be aware of all the resources in the community, particularly in a large metropolitan area. In that case, he should cooperate with a social worker who has this background or with an appropriate agency interested in the problems of handicapped children. These people can serve as sources of information to extend his own knowledge.

Any physician who provides ongoing care for a neurologically handicapped child must see to it that he has available all pertinent information about his patient. This means that he must be sure to receive reports about all visits to medical specialists. He should insist on periodic reports from the child's physical, occupational, and speech therapists. He should request frequent reports from the child's school or educational program. Families should be willing to have this information released to their doctor on a continuing basis. They should also expect that any physician who is central to the long-term management of their child's neurological problems will read this information and not merely file it away.

It is important that the family physician have this information, because he may have to explain and enlarge upon the opinions of medical specialists with whom the family may not have had the opportunity to talk at length. The family physician may have to reinforce, update, or revise information that has been provided in the past. At times, a child may see several medical specialists who have different opinions. For example, an orthopedist may feel that surgery is necessary to modify a handicap; a neurologist may disagree, arguing that surgery will not alter the ultimate outcome of the disease. The physician who has the central role in the child's care is expected to explain the basis of these differences and help parents choose the course of action that is best for their child. During the long-term care of a handicapped child, parents should be able to rely on their family physician to suggest or approve changes in the program of physical or occupational therapy or to write prescriptions for mechanical aids that the child may need. The family should be able to call on him to explain to educators how the child's condition affects his ability to participate in school programs. For example, a child with seizures should not be asked to climb a rope in gym. At times, the doctor may even be asked to mediate between parents' ideas about their child's education and the ideas of those in the public school system. It may be necessary that he find an alternative to the public school if the classroom situation that is available does not meet the child's needs (see Chapter 7).

At times, a neurological handicap may be so severe that it cannot be handled at home without danger to the patient or disruption of the entire family. The family physician who helps care for children with handicaps must be aware of this possibility and prepare a family for institutionalization of their child if he feels that it will be necessary. Placing a handicapped child in an institution typically provokes great guilt on the part of the parents. The physician should be sensitive to the family's need for support at this time. If the physician has treated the child from birth and established a trusting relationship with the family, he will have anticipated the problem long before institutionalization is required and lessened the emotional impact by introducing the idea gradually. This will allow the family to work through their feelings of guilt and accept placement more easily. Under these circumstances, a family at least has the idea that institutionalization can be an appropriate alternative to caring for a severely handicapped child at home. Families should be encouraged to investigate the institutions that specialize in the care of handicapped children before they commit themselves to placement, and they may even become involved in the care given by those institutions by participating as volunteers before placing their child. They can talk to other parents who have had to make such a decision. These preparations, arranged by the physician or his associates, can help assure a family that they are doing what is best for their child.

REFERENCES

Covert, C. 1965. Mental retardation: A handbook for the primary physician. JAMA 191:1–141.

Cunningham, C., and P. Sloper. 1978. Helping your handicapped baby. Souvenir Press, London, p. 333.

Perrin, J. C. S., E. L. Rusch, J. L. Pray, G. F. Wright, and G. S. Bartlett. 1972. Evaluation of a ten-year experience in a comprehensive care program for handicapped children. Pediatrics, 50:793–800.

Shonroff, J. P., P. H. Dworkin, A. Leviton, and M. D. Levine. 1979. Primary care approaches to developmental disabilities. Pediatrics, 65:506–514.

Stewart, M. A., and A. Gath. 1978. Psychological disorders of children: A handbook for primary care physicians. William and Wilkins Co., Baltimore, p. 167.

Talbot, N. B., J. Kagan, and L. Eisenberg. 1971. Behavioral science in pediatric medicine. W. B. Saunders Co., Philadelphia, p. 467.

4

THE MEDICAL SPECIALIST

Many children with neurological handicaps visit a medical or surgical specialist at some time during their life. This contact may be limited to a single visit or at most two or three to advise the family physician or to treat a specific complication. A specialist may also become the doctor who is primarily responsible for the child's care.

WHY GO TO A SPECIALIST?

There are several reasons for limited contact with a specialist:

1. The family or the family physician wants a second opinion about the child's diagnosis, prognosis, or treatment.

2. The family physician does not think he is equipped to provide genetic counseling (e.g., whether their child's disease could be inherited), or he feels unable to evaluate other members of the family to be sure they cannot transmit the disorder to their children.

3. The child requires special surgical care. In the case of neurologically handicapped children, this usually means orthopedic or urological surgery to relieve an acute problem or to avoid the development of a chronic one.

4. Another professional, generally a teacher, physical or occupational therapist, or psychologist, suggests that the child be seen by a specialist because they have noted some new behavior that worries them. They should explain this behavior to the family in detail and write a descriptive note to the doctor.

In each of these instances, the referral is either for a diagnostic consultation or the treatment of a specific, well-defined problem. The medical specialist may see the child once or twice, or he may wish to follow the child periodically to confirm his diagnosis or gauge the effects of treatment.

There are also a number of reasons for continued contact with a medical specialist:

1. The family prefers that their child's neurological handicap be cared for by a specialist rather than by the family physician.

2. The family physician feels that he has neither the time nor the expertise to treat the child's neurological problems.

3. The child's problem is extremely complex and will require the use of sophisticated procedures to establish the cause of the disorder and treat it. These may be available only at a major medical center where physicians have experience treating rare diseases.

4. The child is on a public aid program that demands that the continuing care of a neurological handicap be in the hands of a specialist.

A medical specialist will rarely care for the general pediatric problems of the child, such as an earache, sore throat, or measles. Thus, the family should expect that he will be in regular contact with their family physician, since the nature of the child's neurological problem and its treatment may affect treatment of these medical disorders.

However, the child who is receiving continuing care for a neurological handicap from a specialist is seen as often as necessary. This may be as seldom as once a year or as often as once a week. How often the child is seen is not crucial. What is important is that both the family physician and the specialist agree that the primary care of the neurological problem is the responsibility of the specialist. The frequency of the child's visits to the specialist will depend on the nature of the illness and can be expected to vary considerably during the child's development. A child whose seizures are well controlled may have to see a neurologist only once or twice a year. On the other hand, if his seizures occur more often and can no longer be easily controlled, his visits may need to be much more frequent.

Distance will also help determine the frequency of visits to a specialist. A child who lives several hundred miles away may not be able to see a specialist frequently without disrupting family life. In such a case, the specialist must make the best possible use of resources in the family's community, and he or an associate must be available to discuss problems with the child's family by telephone. Many specialists do not like to make decisions about therapy on the basis of telephone conversations with the family without having recently examined the child; the chance for error is certainly increased. Physicians know this and understand that they are as legally and morally responsible for advice they give by telephone as they are for that given in their office or in the hospital. On the other hand, judicious use of the telephone may be the only way in which a specialist can provide continuing medical care for a child who otherwise would not receive needed attention. A specialist must know how to use the telephone to the advantage of the patient.

The physician who uses the telephone for continuing care should be familiar with local resources so that he can direct parents to the appropriate facility when the need arises. He must have the cooperation of the family physician and the local hospital. Routine laboratory studies needed to make an intelligent decision by telephone can be obtained in a hospital near the patient and the results phoned to the specialist. Blood and urine samples for special laboratory studies that cannot be performed locally can frequently be mailed to a medical center. If the family or the neurologist concludes on the basis of their telephone conversation that an emergency exists, the specialist can ask that the child be brought to his family physician or to the local emergency room. The specialist can then discuss the matter by phone with the doctor who has just evaluated the child.

Any physician who assumes the responsibility for the continuing care of a child's neurological handicap, including the medical specialist, should be kept informed about the child's progress in physical and occupational therapy and about his school performance. The results of any psychological evaluations that are done should be reported to him promptly. He should also be kept advised by the family physician of any unusual changes in the child's general health or his neurological problem. The specialist should keep the family physician, the school, and the child's physical and occupational therapist informed of the child's medical status and particular needs after each office visit or hospitalization. The family of a handicapped child should not be surprised or upset if they are asked to sign forms allowing information about their child to be circulated among the professionals who are caring for him. In fact, it is helpful for a family to have copies of all medical, psychological, therapy, and educational reports about their child in their own files. Parents have every right to request these reports, and they are often a great help if the family moves, goes on vacation, or eventually chooses to see another doctor.

Most specialists who see a child in their private practice require that a referral be made by another physician. However, specialty clinics almost always will accept referrals from public agencies. Some specialists will also accept direct referrals from physical and occupational therapists, psychologists, a school nurse, or a guidance counselor. In our opinion, this kind of direct referral often leads to inappropriate use of the specialist's time and unwarranted expense to the family. In some instances, it also leads to less satisfactory medical care, since the family physician, having been bypassed in the process, feels his responsibility to care for the child's neurological problems is limited. If a child has not had contact with a specialist in the past and such a referral is suggested by a professional other than the family physician, we recommend that the parents consult the family physician first. Most family physicians will have had contact with individual specialists in various medical and surgical disciplines, and they will be able to refer a child to a physician whom they consider reliable and with whom they have a positive working relationship.

If there is a disagreement between the family physician and other professionals who are helping to care for a child about the need for a second opinion from a specialist, the family has the right to ask the physician to make such a referral, and he should be prepared to do so. If he refuses to make a referral, the family can use the Dictionary of Medical Specialists, which can be found in most public libraries, to contact a physician who is board-certified in the relevant medical specialty. These specialists are listed geographically, and a family should be able to choose a doctor who practices nearby.

INFORMATION A SPECIALIST CAN PROVIDE

Families ask the specialist many of the same questions they have asked their family physician:

What is wrong with my child?
What caused his problems?
Will they get better or worse?
Is the condition treatable?
Can his disorder be inherited?
Is there any research going on that may help my child in the future?

It is quite possible that the specialist, despite his expertise, may not be able to give the family a specific diagnosis.

We cannot stress too often that words such as epilepsy, seizures, convulsions, mental retardation, borderline intelligence, specific learning disability, hyperactivity, and even cerebral palsy are terms that are used to describe signs or symptoms; they are not diseases. These terms indicate disordered brain function that affects some aspect of control of movement, thinking, learning, or consciousness but they do not indicate *why* such a problem has occurred.

A syndrome is a group of symptoms and signs that have been seen together often enough to allow a physician to say that they are linked and to give them a name. The Rubinstein-Taybi syndrome is an example. In this syndrome, mental retardation is linked with broad thumbs and toes, a downward slant to the eyes, and a high-arched palate. If a child has this syndrome, one can predict that he has about a 20 percent chance of developing seizures later in life. One can also predict that it is very unlikely that any future children in a family will have the same disorder, since it is not inherited according to simple genetic rules. The use of the term "very unlikely" protects the physician legally although it often fails to satisfy the family. Generally, it is a doctor's way of saying the chances of having a second child with a similar problem are less than 1 in 25 or not much more than those of any other family in the general population.

Unlike a symptom or syndrome, a disease has a known cause. Thus, infection of the fetus early in pregnancy by the German measles (rubella) virus may cause severe damage to the child's brain. The damage may result in mental retardation, a small head, signs of cerebral palsy such as spasticity, and by deafness, visual loss, or epilepsy. All of these symptoms and signs are manifestations of a disease, congenital rubella, contracted by the baby *in utero*. By the time the child is born, the disease may have run its course, but the symptoms and signs of the brain damage it caused to the baby remains. Knowledge of the disease responsible for a symptom or a syndrome allows the doctor to predict with more accuracy the susceptibility of future children to similar problems. However, it is not possible to identify the disease(s) responsible for many common neurological symptoms, such as cerebral palsy, and in many instances vigorous attempts to find a "disease" are unnecessary and undesirable. The physician should explain this to the child's parents. He should indicate what tests will be done and why. Usually only those tests are ordered that will better describe the extent of the brain damage or help treat the symptom. Some specialists will also indicate why certain tests will *not* be done. This is generally because any information they might yield would not affect treatment or because the tests would involve some risk to the child. These are judgments that specialists make each day in their area of expertise.

If the specialist's diagnosis is merely a restatement of a symptom in Latin, what is the value of a diagnostic referral? First, the specialist may be able to assure the family that the child does or does not have a neurological abnormality. If there is an abnormality, the specialist should be able to describe it in greater detail than can the family physician. The parents of a handicapped child should expect the specialist to supply them with detailed information about the extent of the child's handicap that will help them understand his present limitations and future potential. This information should also be useful to other professionals who might have to evaluate or treat the child.

Second, a specialist should be able to outline the child's prognosis. If a child is too young for him to give a reliable prognosis, the specialist should be able to indicate at what stage in the child's development he would be willing to make a more definite statement, and which professionals should participate in the child's evaluation in order to give the family the most reliable information.

Third, a specialist may be able to supply information about the parents' chances of having another afflicted child. Some statistical information is available about the recurrence of symptoms in families even if an underlying disease cannot be identified. Many times tests are ordered that will not necessarily benefit the child but that will provide information of use to parents when they decide whether they want more children. Genetic counseling is an important part of a specialist's function. Therefore, it is important that he discuss with parents their thoughts about future children and that he know the ages of the patient's brothers and sisters. If one of them is old enough to marry in the near future, the neurologist may feel it wise to make further attempts at identifying a disease rather than describing a symptom.

Fourth, the specialist should be able to help plan an approach to a child's symptoms that will help him function as normally as possible. This is necessary even if the specialist cannot make a specific diagnosis or if other professionals will be providing the care. For example, if a child has a learning disability, his treatment will have to be coordinated with the school system or with professionals in the community (usually psychologists) interested in treating these disorders. If no treatment is suggested, parents certainly should know at what future time the specialist feels such a program can be planned and why nothing can be done for the child immediately.

If parents have taken their child to a neurologist and they are not satisfied that they have a clear understanding of his condition, they should discuss the matter with their family doctor. It is possible that a family physician can help parents understand their child's disorder by interpreting the medical specialist's report to them.

In general, physicians try to discourage families from seeking opinions from several specialists in the same discipline. If the diagnosis, prognosis, and treatment of the child's disorder have been adequately discussed with the family, there is usually no reason for further opinions. A child has little to gain by parents maximizing small differences in the opinions of several specialists. If a child's parents and his family physician disagree with the opinion of the specialist, however, it may be advisable to seek the advice of another doctor. In addition, there are occasions in which a second opinion by another specialist in the same discipline may be indicated. For example, a child may be diagnosed as having an untreatable, progressive disease that will eventually result in death. Parents usually will want this terrible reality confirmed by another specialist. This is understandable; any specialist or family physician should be prepared to accept the desire for a second opinion under these circumstances. Parents might also wish another opinion if told their child needs extensive reconstructive surgery. At times specialists in different disciplines, e.g., an orthopedist and a pediatric neurologist, may differ in their ideas about the treatment appropriate for a child. If this is so, it is reasonable to seek other opinions.

One of the specialist's obligations is to help families avoid methods of treatment that have either been proved ineffective or that have not been shown to be as effective as claimed. Unfortunately, there are individuals and groups in most communities that are willing to prescribe lengthy and expensive treatment programs involving untested diets, exercise programs or equipment for any child with a neurological handicap regardless of the cause or the severity of the disorder. They lure families into therapy by creating the hope of a "cure" or of a remarkable improvement in the child's handicap. Such people usually take advantage of the fact that most children with neurological handicaps (if they do not have degenerative diseases) will, if given enough time, develop some new skills. If parents are so persistent in seeking help for their handicapped child that their ability to be reasonably critical and objective is impaired, they will certainly find someone who is willing to take their money and institute a vigorous therapeutic program where none may be needed or where it may do real harm. If several specialists advise against such a program after a thorough evaluation of the child, parents would be unwise to assume that their physician's unwillingness to endorse the program is simply an example of the notion that physicians' minds are closed to new ideas. Before a child is enrolled in any program, parents should investigate it thoroughly and talk with other parents whose children have been in the program for some time. They should also ask for the names of parents whose children have left the program so that they can determine for themselves how much benefit has been derived from treatment.

REFERENCES

Bray, P. F. 1969. Neurology in pediatrics. Yearbook Medical Publishers, Chicago, p. 514.

Farmer, T. W., ed. 1975. Pediatric neurology, 2nd ed. Harper & Row, Hagerstown, Md., p. 558.

Gellis, S. S., and M. Feingold. 1968. Atlas of mental retardation syndromes. U.S. Government Printing Office 0–310–072, p. 188.

Holmes, L. B., H. W. Moser, S. Halldorsson, C. Mack, S. S. Pant, and B. Matzilevich. 1972. Mental retardation: An atlas of diseases with associated physical abnormalities. The Macmillan Co., New York, p. 430.

Littlefield, J. W., and J. DeGrouchy, eds. 1978. Birth defects. E.M.I.C.S. series no. 432, Elsevier, Amsterdam, p. 451.

Menkes, J. H. 1980. Textbook of child neurology, 2nd ed. Lea and Febiger, Philadelphia, p. 695.

Smith, D. W. 1970. Recognizable patterns of human malformation: Genetic, embryologic and clinical aspects, 2nd ed. W.B. Saunders Co., Philadelphia, p. 368.

Swaiman, K. F., and F. S. Wright, eds. 1975. The practice of pediatric neurology. The C.V. Mosby Co., St. Louis, p. 1082.

Warkany, J. 1971. Congenital malformations. Yearbook Medical Publishers, Chicago, p. 1309.

5

THE PSYCHOLOGIST: ASSESSMENT

The handicapped child most frequently is asked to visit a psychologist for an assessment of his intellectual level. However, this represents only one of the contributions a psychologist can make to the comprehensive care of the child. Child psychologists offer a variety of services, including diagnostic evaluations; individual, family, and group therapy; counseling; and training in behavior modification.

The purpose of Chapters 5 and 6 is to inform parents about the different services child psychologists can provide and to discuss the ways in which parents can use these services most effectively.

In this chapter, we will look at various aspects of psychological assessment, highlight the historical development of psychological tests, present an overview of the major assessment instruments and discuss how development level, intellectual capacity, academic learning potential and social-emotional-adaptive behaviors are measured. We will also discuss problems encountered in the evaluation of handicapped children, some special tests and modifications in the administration of tests that can be used to minimize these problems. We will consider the kinds of information that should be included in a psychologist's report so that it will be most helpful to parents and to other persons involved in the care and education of the child.

The first part of the next chapter will be devoted to a discussion of the kinds of counseling a psychologist can provide and will identify situations in which he can be of assistance to the family of a handicapped child. A psychologist, for example, can offer parents support in dealing with their feelings about their handicapped child, as well as their feelings about and perceptions of the reactions of siblings, relatives, friends, and people in general toward their child. He can also provide support and guidance to the child, who must deal with his feelings about his handicap and the impact it has on his life.

The second part of the next chapter concerns itself with behavior management. Handicapped children sometimes develop behavioral problems as a part of their disability or as a result of parental lack of knowledge, overprotection, indulgence, permissiveness, or in some cases, parental rejection. Psychologists who offer training in the principles and methods of behavior management can teach parents how to use specific behavior modification techniques to lessen the frequency of maladaptive behaviors and to increase the frequency of appropriate behaviors. By working with the whole family, a psychologist can help all its members learn to interact with one another more effectively, so that the behavioral problems and emotional stresses so often found in families with handicapped children do not develop.

The psychologist should also be able to direct parents to special parent groups and community services. When parents first find that their child has special educational, physical, or emotional needs, they usually lack the information to locate these services. Parents of a handicapped child need advice and direction to find the proper resources to help their child obtain the particular services he needs. Psychologists are generally familiar with community educational and psychiatric programs and can often save parents countless frustrating hours by putting them in contact with the right person in the right program.

It is particularly important for parents of a handicapped child to become well informed about psychological tests, since their child is much more likely to be evaluated at an early age than a "normal" child. He also stands a greater chance of having his educational opportunities determined on the basis of such tests than does the "normal" child. Therefore, we have included a brief discussion of psychological testing in general, how such tests were developed, and a more detailed examination of the psychological testing of handicapped children.

HISTORICAL DEVELOPMENT OF PSYCHOLOGICAL TESTS

Psychological tests as we know them today came about as a result of scientific efforts to understand the nature of "intelligence." The first "mental" test was based on the idea that people obtain all of their knowledge from their senses (seeing, hearing, touch, etc.). Early scientists reasoned that the person whose "senses" were most acute would therefore be the most gifted and the most knowledgeable. As a result, these first "mental" tests were in fact measures of sensory discrimination: visual acuteness, the ability to recognize tones, the ability to see differences in color, and many other sensory functions.

Next came the development of tests that measured specific skills or talents that were thought to make up mental ability. These tests were designed to measure memory, imagination, comprehension, attention, motor skills, and appreciation of art. These measures were more promising, but they still showed a poor relationship to an individual's actual scholastic or vocational performance. Scientists began to recognize that "general intelligence" was best described or characterized as a mixture of abilities rather than as any single ability, and that performance on a number of carefully chosen mental tests would probably be needed to estimate the level of a person's intelligence.

General characteristics

Many psychological tests are designed to make quantitative statements. Like most other measurements, they yield a number that can be used to compare one individual with another. For example, we can measure Mary's height with a yardstick and say "Mary is five feet tall." Using an intelligence test, we can say, "Mary's I.Q. is 100." Numbers allow us to make comparative statements: Mary is shorter than her brother John, who is five feet, three inches tall; or Mary has a higher I.Q. than Sue, whose I.Q. is 75.

While the measures obtained from psychological tests are similar in some respects to other, more familiar measures, they have other characteristics that are quite different. First, unlike measurements from yardsticks, measurements obtained from psychological tests are indirect. A yardstick measures height directly. There is, however, no physical entity called "intelligence." Nothing tangible within the child is measured with a psychological test. These assessments measure the child's performance (i.e., behavior) on a test at a certain point in time. The scores he makes represent his "demonstrated capacity." While many psychological tasks attempt to quantify abilities, the most important result of a psychological assessment may be a nonquantitative subjective opinion by the psychologist. For instance, if the child is handicapped, observing *how* the child works at a test may be more informative to a skilled psychologist than the score the child achieves. The child's problem-solving approach may permit the psychologist to make some educated guesses about the child's "basic potential."

Second, psychological measurements are relative. There are no units that have a specific meaning. In contrast, when one measures heat, each number (degrees centigrade or Fahrenheit) on a thermometer indicates a specific amount of heat, with definite reference to a point such as absolute zero. But intelligence and other psychological tests have no absolute zero. A child who answered too few questions to score on a test would be considered below the test scale. He could not receive a negative score and how many "units" he was below could not be known. The score earned by one child on a psychological test can only be interpreted by comparing his score with the performance of other children on the same test. We can talk about how one child's performance compares with the performance of others on a test. We can talk about how one child's performance differs from the average child's performance, but we cannot say a child has zero intelligence, or that, for example, an I.Q. of 100 represents an "absolute" measure of his ability.

A third characteristic of most tests is that they do not yield a measure of all aspects of an individual's total abilities. Thus, the results of most tests should be thought of as a sampling of a specific and limited part of a child's total performance. For example, even if a child does poorly on a test, he may still be able to play ball or take a puzzle apart and put it back together again better than his friends.

Fourth, other factors that affect test performance, but that are not included in most measures, need to be considered. These may include the child's motivation, the level of his anxiety, the length of his attention span, or the amount of stimulation he is provided at home. Any of these things may contribute to a child's score, but how much they contribute to that score cannot be measured exactly.

Although psychological tests may not reflect a child's abilities precisely, they do have qualities that make them very useful. For example:

1. Intelligence tests can provide an assessment of a child's general ability more quickly and more economically than almost any other procedure.

2. Standardized tests give an objective measure of ability, which is usually more accurate than subjective impressions. There is less opportunity for personal bias in the evaluation. This is particularly important with handicapped, shy, or otherwise different children whose abilities might be underestimated if they were judged by impressions alone.

3. Numerical test scores or developmental age levels determined from reputable psychological tests are more useful and exact than a verbal description. For example, it is useful, in terms of educational planning or for following his developmental progress over a period of

years, to be able to say that four-year-old Jimmy has an I.Q. of 120, a mental age of five years, four months on the Stanford-Binet test, that his reading, spelling, and arithmetic skills are at a midkindergarten level, and that on the basis of these findings, one might recommend that he be placed in an advanced section of a preschool program. Numerical or developmental age scores are particularly important if, as is often the case with handicapped children, they must take part in various treatment regimens over a period of years. Given a relatively exact objective starting point, one can follow a child's progress over time and determine whether he has been helped by the treatment.

The above are some of the useful qualities of most psychological tests, but each test has specific technical characteristics that influence the psychologist's decision to use one test rather than another. These characteristics are generally not discussed in psychological reports, but we feel they should be described briefly so that professionals and parents can better understand the meaning of the results of a child's psychological assessment. A psychologist or school counselor is unlikely to ask a child's parents to decide which tests should be used to evaluate their child. However, if parents are familiar with the names, the background, and the composition (i.e., the kinds of tasks included) of the more commonly used, well-established tests, they will be able to ask the kinds of questions that will permit them to judge whether the evaluation was appropriate and if the results can be used with confidence to guide educational planning. There are different kinds of psychological tests and a wide array of instruments designed to measure specific aspects of functioning. Table 5.1 lists those most commonly used.

SOCIAL ISSUES IN TESTING

Over the years, the use of mental tests and proficiency tests has stimulated a great deal of controversy and confusion. Some people consider tests "unfair" because, in their opinion, the tests put certain children (minorities and the handicapped) at a disadvantage. Such critics have argued that these tests measure only very limited aspects of intelligence, particularly the ability to learn and remember information taught in school, and that they fail to take into consideration differences in background, environment, handicapping conditions, or qualities such as motivation, creativity or special talents. However, tests themselves are neither "fair" nor "unfair." Professionals who give tests may be "unfair" and their inferences may be unjustified,

Table 5.1. Tests commonly used by psychologists working with handicapped children

Type of test	Age range	Type of score	Description of test	Comment
INDIVIDUALIZED INTELLIGENCE TESTS				
Bayley Scale of Infant Dev.	8 wks.–2½ yrs.	Mental Scale (MDI) Motor Scale (PDI)	A relatively new test; has both a motor and social component.	
Cattell Infant Intelligence Scale	8 wks.–2½ yrs.	M.A. and I.Q. (mental age and intelligence quotient)	Taps motor, social and some verbal skills.	
Stanford-Binet Intelligence Scale	2 yrs.–Adult	M.A. and I.Q.	More verbal than the Wechsler tests; yields one global I.Q. score.	
Wechsler Preschool & Primary Scale of Intelligence (WPSSI)	4 yrs.–6½ yrs. Performance I.Q. Full Scale I.Q.	Verbal I.Q.		
Wechsler Intelligence Scale for Children—Revised (WISC-R)	6½ yrs.–16 yrs.	Verbal I.Q. Performance I.Q. Full Scale I.Q.	All three Wechsler Scales involve both performance and verbal sections. Include several subtests that can yield specific information as to a pattern of strengths and weaknesses.	
Wechsler Adult Intelligence Scale (WAIS)	16 yrs.–75 yrs.	Verbal I.Q. Performance I.Q. Full Scale I.Q.		
SPECIAL I.Q. TESTS FOR THE HANDICAPPED				
Hiskey-Nebraska Test of Learning Aptitude	3 yrs.–16 yrs.	Learning Age (L.A.) for deaf M.A. for hearing	For the deaf or speech-handicapped; relies heavily on visual materials.	
Leiter International Performance Scale	3 yrs.–16 yrs.	M.A. and I.Q.	For the deaf or speech-handicapped; relies on visual materials.	
Peabody Picture Vocabulary Test	2½ yrs.–18 yrs.	M.A. and I.Q.	Can be used with language-disabled children or cerebral palsied.	

Test	Age Range	Scores	Description	Comments
Randall-Island Ravens Progressive Matrices	3 yrs.–Adult	Percentile	A series of matrix designs. Subject is required to choose correctly the missing piece of a design from several possibilities.	
ACHIEVEMENT TESTS				
Gray Oral Reading Test	Grades 1–12	Grade equivalent	Series of progressively more difficult passages with questions following.	Can be modified for visually handicapped using large type (both Gray & Jastak).
Jastak Wide Range Achievement	5 yrs.–45 yrs. and over	Grade equivalent Standard score Percentile	Samples spelling, arithmetic, and sight vocabulary.	
Woodcock Reading Mastery Test	Grades K–12	Grade scores Age scores Percentile ranks Standard scores	Five tests that sample letter identification, word identification, word attack, word comprehension, and passage comprehension.	
LANGUAGE ABILITY TESTS				
Illinois Test of Psycholinguistic Ability (ITPA)	2 yrs.–10 yrs.	M.A. Psycholinguistic age (PLA)	Ten subtests used to pinpoint areas of difficulty in communication.	Often used as a screening device to locate learning disabilities.
VISUAL-MOTOR TESTS				
Bender Visual Motor Gestalt Test	5 yrs.–Adult	Test age	Nine simple designs presented one at a time for a subject to copy.	
Benton Revised Visual Retention Test	6 yrs.–13 yrs.	I.Q. range	Ten designs, presented one at a time for several seconds, then removed. Subject to remember and draw them. (Both the Benton and CVRT.)	
Children's Visual Retention Test (CVRT)	3 yrs.–5 yrs.	Number correct		
Draw-A-Person Test	3½ yrs.–Adult	M.A. and I.Q.	Subject is asked to draw a picture of a man, woman and himself.	
Frostig Developmental Test	3 yrs.–8 yrs.	Perceptual quotient	Widely used as a measure of perceptual skills.	

Table 5.1. (Continued)

Type of test	Age range	Type of score	Description of test	Comment
Hidden Figure Test	8 yrs.–Adult	Number correct	Subject required to locate and trace smaller figure embedded in larger one.	
Sequin Form Board Figures	6 and 7 yrs.	Number correct	Eight designs, same rules as Benton & CVRT.	
AUDITORY DISCRIMINATION TEST				
Wepman Auditory Discrimination Test	5 yrs.–Adult	Rating scale range	Child listens to pairs of words and decides if same or different.	
SOCIAL-EMOTIONAL ADAPTATION TEST				
The Piers-Harris Children's Self-Concept Scale	Grades 4–12		A self-report questionnaire dealing with "How I Feel"-type questions.	
Vineland Social Maturity Scale	Below 1 yr.–Adult	Social age and social quotient	A social maturity checklist usually administered via a parent interview.	
SCREENING TESTS				
Denver Developmental Screening Test	1 month–6 yrs.	Normal or abnormal	Used to detect child with significant motor, social, and/or language delays. Information is gathered from observation and from parent questionnaire.	
Developmental Screening Inventory	4 wks.–18 months	Normal–questionable–abnormal	Items tapping five fields of behavior—adaptive, gross-motor, fine-motor, language, personal-social. Assessment made from observation—items presented to child. History—parent response to questions.	

but if reputable tests are selected, administered, and interpreted by well-trained psychologists, the results can accurately identify areas of intellectual strength and weakness and, thus, can be extremely useful in educational and vocational planning.

Tests of general mental ability were formulated to predict what the school performance of children of a similar age would be, based on their ability to answer questions, solve problems, or complete nonverbal assignments of increasing difficulty in a number of different areas. They included samples of nonverbal abilities. These tests make no statement about *why* differences exist between individuals or groups of individuals, nor do they state that future performance might not be modified if the environment were significantly altered.

SELECTING TESTS FOR A CHILD

When selecting a test or group of tests to use to evaluate a child a psychologist should consider certain features about the child and about each test before making a decision to use "Test A" rather than "Test B" or one combination of tests rather than another. These considerations include the "appropriateness" of a specific test for a particular child (the child's age, physical ability, etc.), and whether the test is accepted by reputable psychologists. An equally important factor to be considered is the reason for the evaluation. Thoughtful test selection is particularly important in the case of handicapped children, whose educational and vocational opportunities are very often determined on the basis of data gathered from psychological tests.

Characteristics of the child

Information about the child that determines which test(s) will be used includes the child's age, the reason for the evaluation, the nature and extent of any handicapping condition, and any other special features about the child that might influence his performance on the test. For example, the selection of tests to evaluate a child whose family had just immigrated to this country from Spain would certainly be influenced by the child's ability to speak and understand English. Therefore, it is essential that parents inform the psychologist of any special or limiting characteristics of the child before he is tested.

Characteristics of the test

There are three major test characteristics that most psychologists would agree determine whether they would feel confident about the results it yielded. These three characteristics are the reliability, validity, and the standardization of the test. The term reliability, when used in reference to a psychological test, always means the consistency or stability of the scores the test yields. To be reliable, a test must (1) yield relatively similar scores when used by different psychologists; and (2) yield relatively similar scores for the same child if he is tested at different times.

The reliability of a test is very important, particularly for children who must be reevaluated periodically. Scores on the same or alternate test forms (developed by the same author) are used to monitor a child's reaction to medication and therapy. If the test is reliable, significant changes in a child's scores can be used to alert the physician to the need for a change in therapy. Deterioration in performance may signal deterioration in physical well-being. Using the same test makes it easier to compare performance over time.

Tests such as the Wechsler or the Stanford-Binet Intelligence Scale for Children are well-designed, reliable tests that have been used for years with children from six to sixteen years of age. Using either of these tests, for example, one could reasonably expect that if a child obtained an I.Q. of 110 at age seven, his score would probably not vary more than ten to fifteen points if he was retested with the same test at age ten and again at age sixteen unless some drastic change in the child's situation had occurred.

A number of situational factors may influence reliability, such as the child's physical condition (e.g., whether he is tired, hungry, just getting over a cold, etc.), the person doing the testing (whether he is inexperienced, unfamiliar with the test, tired, etc.), or even the room the child was tested in (whether it is too hot, too cold, poorly lit, etc.). These are a few examples of situational factors that could influence reliability. There are, in addition, complex statistical aspects of reliability that are considered in test construction, but we need not be concerned with these here. It suffices if parents understand reliability to mean that if their child is tested with the same test at different times, and by different psychologists, the results should be pretty much the same unless some major change in the child's condition has taken place. Test reliability is a major concern if test results are used to predict future performance. For example, if the testing is being done for long-range predictions, then the reliability of the test or the stability of the test over a period of years is important.

The second, and probably the most important, factor to be considered when a test is selected is its validity—that is, the degree to which the test measures what it is supposed to measure (e.g., intelligence, reading, spelling, arithmetic, motor abilities, or some specific aspect of a child's abilities). A valid intelligence test, for example, must sample a sufficient number of the different behaviors (i.e., abilities and skills) that most professional and lay people would agree reflect intelligence. A test that was represented as an intelligence test that sampled only the ability to repeat a series of numbers after the examiner would clearly violate the notion of validity. Intelligence, as defined by most people, is the ability to take in information, store it in one's mind, remember those bits of stored information when necessary, combine them with new information, and use this reorganized information to cope with a situation, solve a problem, or understand new experiences. Sampling a single mental ability, i.e., memory, information, puzzle solving, etc., fails to measure intelligence adequately. Testing multiple abilities, as is done in tests such as the WISC-R or Stanford-Binet, is a much more valid measure of the elusive quality of intelligence.

If one is trying to determine in which class at school a child should be placed, then the validity of test scores (whether of intelligence or of specific school or vocational skills) is a more immediate concern than their reliability. The extent to which a test answers the questions being asked is a measure of its validity. For example, a reading test would be considered valid if it included a representative sample of the reading materials a child would have had the same chance to learn as other children his age.

There are several additional aspects of validity that are more technical than we need to discuss here. However, parents concerned about the validity of a test should talk with the psychologist in charge of the evaluation to find out how well it measures what it is supposed to measure. Questions they might want to ask include the following:

1. Which tests were used?
2. What were the test items like (i.e., what things was the child asked to do)?
3. What was the test designed to measure?
4. Does the psychologist feel confident about the results obtained?

A third quality of a test is its "standardization," or how objective the test is. Standardization has to do with the procedures that are followed in the administration and scoring of a test. A well-constructed standardized test minimizes any effect that the examiner's values, biases, and judgments might have on the results, since each child's responses are scored according to the strict rules that govern the procedures used in giving the test. To have confidence in the results of an assessment, one must assume that the person administering the tests is adequately trained and sufficiently skilled to present each test item, record the child's responses, and score the responses according to the instructions set forth in the manual that accompanies each test.

Finally, psychological test scores, as we noted earlier, are not absolute measures; they are meaningful only when they are compared with the scores produced by the group of individuals who were tested with the same test and whose scores were used to develop the "norms" for the test. The techniques involved in the construction of a table of norms for a test need not concern us here, but a few words about "norms" can help us understand how these are established. Tables of "norms" are derived from the scores produced by age-matched sample populations tested with the same instrument. This means that a child's score is compared with the scores achieved by the "sample" group of children of his age whose scores on the same test were used to make up the "norm" table. The scores are generally reported as intelligence quotients (I.Q.s), percentiles, or standard scores. Each of these methods of reporting age norms allows the psychologist to compare one child's performance with the average performance of others his age. It is this particular quality of psychological tests that presents the greatest problem when the performance of a handicapped child is evaluated, since most "norms" are derived from populations of children or adults with no evidence of physical disabilities.

TESTS OF INTELLIGENCE

Stanford-Binet Intelligence Scale (Fig. 5.1)

The first satisfactory individual intelligence test was developed by Alfred Binet, a French scientist who in 1904 was commissioned by the Minister of Public Instruction in Paris to construct a test for the measurement of the intelligence of Parisian schoolchildren. Paris school officials, alarmed by the number of children who had failed to learn, decided to remove the seriously retarded children to schools in which they could be offered a simplified program of learning. Not wanting to remove children who had good potential, even if they were troublemakers whom the teacher might wish to be rid of, school officials felt they could not trust teachers to identify only truly retarded children. Moreover, they wanted to identify all retarded children, even if they came from good families. They feared teachers might hesitate to point out these children as dull. The school officials felt that if a "mental ability" test could be constructed that would identify those children

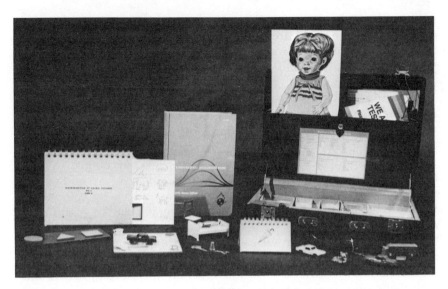

Figure 5.1. Test materials employed in administering the Stanford-Binet Intelligence Scale. Illustrated are a number of the toy objects (e.g., cars, key, kitten, etc.) the children are asked to identify, demonstrate their use, or use in following commands. Also shown are the test manual, the picture vocabulary booklet that contains drawings of objects the child must name, the card used to tap a child's ability to recognize two identical animals, and a picture of a doll used to determine whether a child can point to body parts correctly. The use of these items and others gives the examiner an opportunity to observe a child's behavior and response pattern in the testing situation. (Copyright © 1978 by Warren Paris. Reprinted by permission of Houghton Mifflin Company. All rights reserved.)

who could not profit from the regular classroom curriculum, their problems would be solved. Binet was commissioned to develop a test that could be used to distinguish the lazy or poorly adjusted child from the child who lacked the basic capacity to learn.

Binet, in collaboration with Dr. Théodore Simon, another French investigator, completed the Binet-Simon Intelligence Test in 1905. The test focused on the child's ability to understand and reason. The ability to reason about objects and events in one's cultural environment was considered evidence of intelligence. Consequently, the test items included naming objects, comprehending of questions, and completing sentences. The range of questions included very simple ones that even the slowest child might answer correctly, and difficult questions that would indicate a superior level of ability. Improvements and revisions were made as practical work with these tests revealed weaknesses. Binet's basic assumption was that a child was "normal" if he could do the things children of his age normally do; he was retarded if his test performance was similar to the performance of children younger than himself, and accelerated if his performance excelled that of children his own age.

In 1916, the Terman-Stanford revision and extension of Binet-Simon Intelligence Scale was published in the United States. It represented an attempt to provide standards of intellectual performance for average American-born white children from three years of age to young adulthood. Test questions, or "items," were arranged in order of difficulty by age levels. The intellectual ability of a child, determined by his performance on the Scale, was judged by comparison with the standards of performance for normal children of different ages. Intelligence ratings were expressed as "mental age" scores. A child's relative position in his own age group, that is, whether he was average, brighter than average, or duller than average, was indicated by calculating the ratio of the child's mental age score to his chronological age (the mental age divided by the chronological age and multiplied by 100). This procedure yielded a "ratio I.Q.," which was first employed in the 1916 scale. The 1916 scale was standardized on a sample of approximately a thousand children and four hundred adults. Careful attention was given to the selection of these children and adults in an effort to obtain a representative sample of the general American population. This effort to obtain a representative sample of the population, the development of detailed instructions for administering and scoring each test, and the introduction of the "I.Q." all constituted major pioneering steps in test construction.

The Stanford-Binet was revised in 1960, restandardized in 1970, and revised again in 1973. Test scores of a representative sample of minority groups (based on the United States Census) were included in the 1970 restandardization.

The Stanford-Binet is still considered one of the two most efficiently designed individually administered intelligence tests for children. The test items are interesting to children, the administration procedures are clear, and the validity, reliability, and standardization of the test are well established (Fig. 5.1).

Wechsler Intelligence Scales

There are three major variants of the Wechsler test, each one designed to test populations of different ages:

1. Wechsler Preschool and Primary Scale of Intelligence (WPPSI)— Ages 4 to 6½

Figure 5.2. Materials used with the Wechsler Intelligence Scale for Children-Revised. Included in the display are the test manual, copies of the test forms, a part of the Arithmetic, Picture Completion, and Block Design subtests. Also shown are two demonstration items: one, a puzzle from the Object Assembly subtest; the other, a three card item from the Picture Arrangement subtest. (Copyright 1974 by the Psychological Corporation. Reproduced by special permission from the publisher.)

2. Wechsler Intelligence Scale for Children-Revised (WISC-R)—Ages 6 to 16 (Fig. 5.2)

3. Wechsler Adult Intelligence Scale (WAIS)—Ages 16 to adult

The Wechsler Intelligence Test Series, which evolved from the older Wechsler-Bellevue Test for adults, are individually administered general "mental ability" tests. Wechsler, like Binet, viewed "intelligence" as a complex function, rather than a single ability. He, therefore, followed Binet's lead and used a combination of tasks, each sampling a different kind of behavior (ability) to arrive at a measure of "intelligence." However, instead of mixing different kinds of tasks according to their difficulty, Wechsler arranged items into subtests by content; thus, all vocabulary items are grouped together in one subtest, all puzzle assembly items are offered in one subtest, etc. There are twelve subtests, grouped into a Verbal and a Performance Scale. Within each subtest, items are arranged in order of difficulty, so that each child has an opportunity to accomplish increasingly difficult tasks in each of the subtests. The Verbal Scale includes tests of Information, Comprehension, Arithmetic, Similarities, Vocabulary, and Digit Span (optional). The Performance Scale (nonverbal) includes Picture Completion, Picture Arrangement, Block Design, Object Assembly, Coding, and Mazes (optional). Each individual being tested is given all of the subtests, so that a consistent sample of abilities is achieved. The test yields three I.Q.s—one derived from the Verbal Scale, one from the Performance Scale, and one a Full Scale I.Q., which is derived from a combination of the two scores (Table 5.2).

Wechsler, criticizing Binet's original mental-age concept, introduced standard-score I.Q.s, a scoring procedure that has since been adopted in the Stanford-Binet Tests. Standard Score I.Q.s are figured by changing the actual raw scores (the number of correct answers on each subtest) to a Standard Score provided in tables that were developed from the sample population (one chosen to represent the whole population). The Standard Score is based on the number of correct

Table 5.2. Intelligence classifications [a,b]

I.Q.	Classification	Percent included	
		Theoretical normal curve	Actual sample [c]
130 and above	Very superior	2.2	2.3
120 to 129	Superior	6.7	7.4
110 to 119	High average (bright)	16.1	16.5
90 to 109	Average	50.0	49.4
80 to 89	Low average (dull)	16.1	16.2
70 to 79	Borderline	6.7	6.0
50 to 69	Mentally deficient (educable)		
25 to 49	Mentally deficient (trainable)	2.2	2.2
Below 25	Mentally deficient (profound)		

[a] Reproduced from the Wechsler Intelligence Scale for Children-Revised. Copyright 1974 by The Psychological Corporation. Reproduced by special permission from the publisher.
[b] The President's Committee on Mental Retardation (1976) accepts the definition of the American Association on Mental Deficiency (AAMD) which states "Mental retardation refers to *significantly* subaverage general intellectual functioning existing concurrently with deficits in adaptive behavior, and manifested during the development period." While the AAMD emphasizes adaptive behavior most children are assigned to classes for the trainable or educable retarded on the basis of I.Q.
[c] The percentages shown are for Full Scale I.Q. and are based on the total standardization sample (N = 2200).

answers given by children at each age level in the sample population. For example, a bright, six-year-old who answered 8 questions correctly would earn a raw score of 8 on the Information subtest of the WISC-R and a Standard Score of 14, whereas a ten-year-old who answered the same number of questions on the same subtest would get the same raw score of 8, but only a Standard Score of 4, indicating that the child functions below age level on that particular subtest. As a general rule, only transformed Standard Scores are stated in the psychologist's report. This makes the actual change in a child's abilities over a period of time difficult to evaluate, since a child may have actually gained some new information between tests, but if it was less than his peers, his Standard Score would drop. A comparison of the raw scores allows a psychologist to determine whether a child has developed new skills or added to his fund of knowledge. Therefore, if the results of psychological tests are to be forwarded to another psychologist, parents should ask that all the raw data be included, not just a summary of the Standard Scores and the resulting I.Q. A detailed analysis of the raw data from several tests administered over a period of years would allow a psychologist to tell parents more about their child's abilities and potential than he could by comparing I.Q. scores.

The organization of the Wechsler tests makes them particularly useful in the assessment of handicapped children. For example, the Verbal Scale alone can be used for a child who has impaired motor coordination without distorting the meaning of the obtained I.Q., since it is derived solely from the verbal subtests. Similarly the WISC-R Performance Scale alone can be used for a mute child or a child with seriously impaired speech.

The WISC-R is a sound, well-standardized test which, because of its design (separate subtests, separate scales) and scoring system (separate I.Q.'s) is probably the most preferred test for children between the ages of six and sixteen (Fig. 5.2).

INFANT AND PRESCHOOL TESTS

To use the term "intelligence test" to describe a test for infants below eighteen months of age is misleading because most infants have not yet developed the behaviors usually associated with "intelligence" and measured with intelligence tests. The intelligence tests discussed earlier present specific questions for the child to answer and set tasks for the child to perform, which the examiner can then score according to the rules of the test. This method cannot be followed with infants because they have little, if any, speech, they generally do not understand instructions or follow directions, nor do they consistently do things on command. Much of the controversy about the value of infant testing was stimulated by overly enthusiastic attempts to predict the learning ability and the intelligence of school-age children from scores earned on infant tests. Investigations into the accuracy of these early predictions showed that the tests were poor predictors of intellectual potential. We have since come to recognize that those aspects of the learning process that we can measure change as the child matures, and that intelligence is neither static nor unchangeable. Infant tests are now more properly described as "developmental schedules" rather than intelligence tests. The basic aim of most of these scales, such as the Cattell or the Bayley (Table 5.1), is to determine whether the infant's development is following the expected or "normal" time schedule. Developmental schedules generally cover four major areas:

1. *Motor development:* gross and fine motor abilities (head control, rolling over, sitting, standing, walking, stair climbing, manipulation of objects, raking versus pincer grasp, finger dexterity etc.).

2. *Adaptive development:* reaches for objects, transfers objects from hand to hand, reacts to stimuli, pushes objects away, changes position to reach a desired goal, etc.

3. *Language development:* babbles, coos, vocalizes, laughs aloud, smiles, says "mommy," "dada," makes m-m-m sounds, etc.

4. *Personal social development:* anticipates feeding, watches moving person, hand play, spontaneous vocal-social response, discriminates strangers, etc.

The child's performance on these scales is expressed in terms of developmental quotients that are determined from comparison of the infant's observed patterns of behavior in each of these areas with "normal" behavior patterns. Some infant developmental scales yield "Developmental Quotients" for each of the separate areas tested. This permits one to determine whether there are significant lags in specific areas of the child's development (Fig. 5.3).

DEVELOPMENTAL ASSESSMENT OF NEUROLOGICALLY IMPAIRED INFANTS

In the last decade, research has suggested that neurologically handicapped infants who live in an intellectually stimulating home environment suffer less developmental delay than those who are deprived of

Figure 5.3. Materials used in the administration of the Bayley Scales of Infant Development. Shown are the toys, formboards, picture books, pegboards, and other test objects used to assess an infant's developmental level. (Copyright 1969 by the Psychological Corporation. Reproduced by special permission from the publisher.)

such stimulation. Findings from several remediation programs for children with cultural deprivation, Down's syndrome, sensory handicaps, and other developmental disorders also suggest that such programs significantly improve children's capabilities as measured by standard developmental and psychological tests. These findings have stimulated interest in early detection of developmental disorders and early treatment of any possible delay. The focus of infant evaluation has shifted from prediction of future school performance to assessment of present levels of functioning and identification of problem areas in order to plan a treatment program that will minimize the infant's handicap.

Developmental schedules such as the Bayley (Fig. 5.3) or the Cattell, particularly when combined with information gathered from structured interviews such as the Vineland Social Maturity Scale, are of value in identifying children who are at high risk for developmental difficulties. The Vineland Social Maturity Scale provides a useful measure of a child's social development, self-help skills, and gross motor coordination. The scale is made up of questions related to the social and self-help activities the child engages in at home. The psychologist uses the scale to interview the mother or other caretaking adult about the child's behavior, beginning with activities that should have already occurred before the child's current age, and continuing with questions about current behaviors. The inquiry stops when the parent reports that the child does none of the activities at a particular age level. A developmental quotient can then be calculated based on the parent's answers.

Evaluation of an infant with serious neuromuscular problems or significant vision or hearing defects is generally limited to observations of the child's spontaneous activity and information gathered from interview schedules such as the Vineland. The initial scores are probably not as important as any change in the child's scores over a period of time. All infants identified as developmentally delayed must be followed carefully and reevaluated at regular intervals so that treatment plans can be modified to fit their changing needs. If intervention strategies are initiated, parents should keep a record of their child's progress so that benefits can be objectively determined. Infant stimulation and early intervention programs have become big business; objective evaluations of their results, however, are all too infrequent.

Before a child is enrolled in an early childhood stimulation or remediation program, we suggest that the parents visit the program; observe the procedures used; talk with the teachers and director; and inquire about the goals of the program, the goals for the child, how they will be selected and implemented, and how the effectiveness of the program will be evaluated. Even programs that bear the same name may be offered by different agencies or private nursery schools and may differ widely in procedure and in results. Some programs provide no more than glorified babysitting and others include everything from social competency training to creative thinking and enjoyment of fantasy. The program selected should depend on the goals that are set for the child by the parents in cooperation with medical professionals. This goal should be explicit and seem reasonable, based on the child's present abilities. The continued participation of a child in such a program should depend on how the objectives of the program are implemented and how the success or failure of the treatment is evaluated.

EDUCATIONAL ACHIEVEMENT TESTS

Educational achievement tests measure proficiency in important school-related skills (e.g., reading, spelling). They are most often given to a group (such as an entire class) rather than to each child individually. Although some educational achievement tests combine a measure of more than one skill and test different aspects of each skill (e.g., reading fluency, phonics, reading comprehension), they do not measure general intelligence. Educational achievement tests are used to help teachers and administrators monitor students' progress and make decisions about their grade placement. A student's performance on a test of reading proficiency, for example, would indicate whether the student meets the standards expected for his current grade and also allows teachers to decide whether he is ready to proceed to the next higher grade level. Standardized achievement tests are designed to reduce teachers' bias so that similar scores earned by different students or students in different classrooms will represent a similar degree of proficiency. Ideally all schools should use the same achievement tests so that transfer of students from one school to the appropriate grade level in another would be accomplished with little difficulty. Unfortunately, individual school districts and sometimes even schools within a district use different tests and set different time periods for the evaluation of students' progress. As a result, children can rarely be placed in a new school without further evaluation.

Achievement tests are judged by their "content validity," or how well the test covers the subject matter. Content validity is usually established by logical examination of the test questions. Scores on achievement tests are most often reported in "grade equivalents" (E.Q.). Thus an arithmetic test for a first-grade student would cover all of the number skills that had been taught in class in a certain period. A grade equivalent of 4.6 indicates that the child performs at the level of the average fourth-grader after two-thirds of the school year has been completed. Grade norms are based on samples of students across the country. Obviously, there are differences in the quality of teachers, in the amount of money a given school district has to spend, and in student ability. To compare a handicapped student from a poor school district with a "normal" student from a prosperous school district is unjust. Students should be compared with other students who have similar problems and are in similar school situations. The goal of educational testing of handicapped children should be to evaluate performance relative to ability and to identify gains in performance after a course of instruction.

Educational achievement tests for handicapped children

Special equipment, materials, or techniques should be available as needed to fit the particular problems of the handicapped child. For example, a test normally given against a time limit (e.g., 30 minutes to complete) could, in the case of a moderately spastic child, be given without the time constraints, and the child allowed to continue working until he either completes the test or reaches the point at which the demands are beyond his ability. This procedure gives a better picture of the child's educational achievement than would one with a strict time limit. Spelling tests for a child unable to control a pencil may have to be modified to a multiple-choice form, with the child selecting the correctly spelled form of the word from a number of choices. These modifications must, of course, be taken into account not only when the child's test performance is interpreted in terms of the norms that are provided, but also when a recommendation regarding educational placement is made. For example, it would be grossly unfair to place a child in a regular classroom on the basis of the results of modified tests without insuring that similar modifications could be and would be adopted in the classroom. As long as the methods used to obtain the results are clearly stated, modifications in administration or materials are justified and the results quite valuable. If a physically handicapped child is evaluated for educational placement, his parents are urged to inquire whether the tests were modified to fit their child's needs. If this was not done, the parents should not hesitate to ask what provisions will be made to ensure that the results are a reflection of the child's capacity to learn and not an estimate of the speed with which he can complete assigned tasks.

Educational achievement tests are included in psychological assessment not only to find out what a child has learned in school, but also to establish at what level educators must work with the child in each of the subjects measured. When large discrepancies between achievement and general mental ability (as measured, for example, by the WISC) are revealed, the psychologist may, from an analysis of the child's performance on other tests such as visuo-spatial-motor organizational tests or measures of social-emotional adjustment, be able to identify problems that interfere with his learning. It is also not unusual to find children who do poorly in school yet score at or above grade level when given individual educational achievement tests. This type of discrepancy also suggests that social and emotional adjustment at school should be evaluated for possible problems.

DEVELOPMENTAL SCREENING TESTS

The rapidly growing numbers of government-supported remedial programs for culturally deprived, developmentally delayed, neurologically impaired, mentally retarded, or otherwise handicapped children has led to an equally rapid proliferation of screening programs set up to identify children with developmental problems. Developmental centers and other services of the Bureau of Developmental Disabilities under Public Law 94-103, and the Individualized Education and Therapy Program directed by the Bureau of Education for the Handicapped (Public Law 94-142) must make available assessments of all children suspected of developmental handicaps. The need to screen large numbers of children quickly and economically stimulated a demand for screening tests that are not only quick and easy to administer, but that do not require advanced professional training to interpret. Developmental screening tests by definition are devices used for detecting developmental delays in infancy and the preschool years. They are not meant to be used diagnostically nor are they designed for precise appraisals of specific functions. Unfortunately, some programs use screening tests diagnostically, even though their standardization, reliability, and validity data are limited. The lack of such data about the tests makes interpretation of their results impossible for a layman and difficult even for a professional.

Screening tests are not precise, and often it is not clear how their items relate to the areas of development they purport to measure. Probably the most serious concern about the use of screening tests relates to the manner in which their findings are expressed. A brief screening test is able to assess only very limited aspects of development and that with limited precision. Too often, however, the results are reported to parents as a global assessment of a child's development with little regard for the fact that language, motor, and social abilities do not develop at the same rate. Such misuse of screening tests or procedures must inevitably result in inappropriate labeling that could have an unfortunate impact on a child's entire future. The purpose of a screening test is to separate children who are apparently normal in the area being examined from those needing further study. Therefore, these tests should be used only if provision is made to promptly refer those who need further investigation to an appropriate professional. If this provision is not made, needed therapy may be delayed beyond the critical time.

Screening tests are particularly difficult to adapt to the child with neuromuscular or sensory handicaps, because they do not cover a sufficiently wide range of behaviors to permit accurate assessment of areas of potential strength. Well-designed screening tests such as the Denver Developmental Test, when properly administered and interpreted, do have an appropriate place in the early identification of children suspected of developmental delay. Indeed, the use of such tests may be necessary if professionally trained personnel are unavailable. However, it is important for parents to bear in mind that if they have serious concerns about their child's developmental progress, they should consult with their pediatrician about obtaining a professional psychological evaluation. If the results of a screening test are said to indicate an abnormality, that assessment should not plague a child throughout his early years. Parents should demand a more extensive evaluation and be sure treatment, if indicated, is begun early and is related to deficits that are clearly apparent on validated tests.

There are far too many screening tests currently available for us to review each individually. However, three tests that, in our opinion, deserve mention are the Denver Developmental Screening Test (DDST) (by William Frankenburg and Joseph B. Dodds), the Developmental Screening Inventory (DSI) (by Hilda Knoblock and Benjamin Pasamanick), and the Neonatal Assessment Scale (by T. Barry Brazelton).

THE ASSESSMENT OF SPECIAL GROUPS

Although it is fairly easy to select a group of tests that are valid and reliable for normal children, it is much more difficult to select tests for children who have neurological impairments. Virtually all of the tests commonly used to obtain a psychological profile or description of a neurologically impaired child were designed and standardized on essentially "normal" populations of children. With the exception of mentally retarded (but otherwise normal) children, this raises the question of whether the test performance of a handicapped child should be recorded in terms of scores derived from the population of children used to standardize the tests and to provide the reliability and validity data. Should, for example, the scores of a cerebral palsied child or a child with a serious speech impairment be compared with the scores earned by a normal child? If so, the impaired child must suffer by comparison. On the other hand, if we use tests developed specifically for handicapped children, and if these tests are standardized on groups of children with the same handicapping conditions, we are left with the problem of determining from the results where and how a child will function in the larger society.

Although these questions have yet to be answered satisfactorily, methods have been developed to make the psychological assessment of handicapped children possible. These methods include adaptations in the testing procedures; modification of the test itself or of the materials used; and greater dependence on observations and clinical judgment of *how* the child performs on the test (his motivation, persistence, problem-solving ability etc.), rather than of how *much* he accomplishes. Adapting the testing procedures, but keeping the content of the original test the same, is essentially a matter of the psychologist's using special tactics in the test situation. For example, the examiner may read the standardized items to a partially sighted child or allow a child with a major speech disorder to point to one of several pictured responses rather than replying orally. Removal of time limits is another alteration of standard administration, though this too leaves the content of the test intact. Such alterations test the child's understanding while partially bypassing his specific handicap. In other situations, adaptations involve modification of the original test, either by omission of certain items or by using only those parts of the test to which the child will be able to respond in an acceptable fashion. The purpose of adaptations is to try to sample as much behavior as possible in terms of perception, reasoning, and potential learning disability. Standard administration of many nonverbal tests includes time constraints and bonuses for speed. In such situations, the psychologist should pay more attention to how the child reasoned and went about solving problems than to the length of time it took him to complete a task. The goal is always to obtain a meaningful psychological profile of the child, not to obtain a meaningless numerical score.

In addition to adapting the administration procedures and test structure, a clinical psychologist may exercise some liberty when drawing inferences from the test performance of a handicapped child to arrive at a reasonable estimate of the child's basic capacity. A normal child's test scores are derived from his actual or demonstrated performance, i.e., exactly what he said or did in a standardized test situation. The numerical representation of this capacity is generally unaffected by any explanation of factors that influenced the child's performance. Inferences or assumptions about how much better or worse the child might have done are limited and demand considerable supporting evidence. In working with handicapped children, the clinician will make inferences regarding the child's "basic" capacity from observations of his approach to problems. In some instances, the psychologist emphasizes certain parts of the test and not others to infer a child's basic capacity. For example, a stronger intellectual potential might be assumed for a child whose vocabulary was better than his ability to repeat digits. In other instances, the inferences as to basic capacity are derived from the quality of performance, rather than from the quantity of performance. Obviously, the chances for and the size of the errors in such inferences about basic capacity are much greater than if standardized procedures for test administration are followed. Therefore, it is important for parents of handicapped children to make certain that their children are evaluated by experienced psychologists who understand how to modify assessment procedures to fit the needs of a handicapped child and how to interpret the findings in terms of the modifications.

The hearing-impaired child

Severe impairments in hearing greatly limit the choice of tests that can be used to measure the child's intelligence. A common practice is to use the nonverbal sections of the WISC-R. This involves giving the child the necessary instructions in pantomime. Since this test was developed for normal children, the deaf child could be penalized if the examiner's pantomime skills are poor. If a child's scores are low, interpretation of the results is difficult and frequently inaccurate. If he scores in the average range, it is unlikely that he is retarded since he managed to perform adequately in spite of the difficulty imposed by alteration of the test procedures, but the child's real potential is unknown. A high score suggests that the child is exceptionally bright and probably will be able to learn without difficulty.

There are two tests for deaf or hearing-impaired children that are well accepted among clinicians, though they are limited in their diagnostic value. Probably the best of the tests for deaf children is the Hiskey-Nebraska Test of Learning Aptitude. This test was developed and standardized on deaf and hard-of-hearing children. It is an individual test suitable for children from three to sixteen years of age. None of the tasks included in the test are timed tests, since it is difficult to convey the idea of speed to young deaf children. This test does do a better job of sampling a wider variety of intellectual functions than the nonverbal portions of most standard intelligence tests. The instructions are given in pantomime and through the use of practice exercises. All the items included in the test were chosen with reference to the special limitations of deaf children.

Another test often used is the Leiter International Performance Scale. Although this test was initially developed for use with different ethnic groups, it has proved to be a valuable test to use with deaf children because of the almost complete elimination of either spoken or pantomimed instructions. Each section of the test begins with a very easy task of the type that will be offered in the particular section. This procedure is followed throughout the whole test. The child's ability to understand the task is part of the test. It is administered individually with no time limits, and the tests are arranged into year levels, from two to eighteen.

As noted earlier, there are special adaptations of other standard intelligence tests that are sometimes employed in testing deaf children. For example, if the child is able to read at fifth-grade level, the verbal tests included in the Wechsler Scales can be administered by duplicating in printed form all oral instructions or questions. However, it is important that parents understand that such modifications of standard testing procedures do change the psychologist's confidence in interpreting the results.

The partially sighted or blind child

The problems encountered in testing blind or partially sighted children are quite different from those encountered with deaf children. Verbal tests can be most readily adapted for blind persons because instructions or questions can be presented orally without gross distortion of the meaning of the scores. Some tests, such as the Stanford-Binet, are also available in Braille. However, this technique is somewhat limited in its applicability because of the slower reading rate for Braille and because few blind children are good Braille readers. Both the Stanford-Binet and the Wechsler Intelligence Scale for Children have been adapted for blind persons who are unable to read Braille. For the most part, these adaptations consist of omission of those portions of the test that require sight and substitution of oral for printed material whenever possible. Materials that can be reproduced in large-scale type are used with those children considered "legally blind" who, nevertheless, have sufficient sight to use such materials. Many of the educational materials used in schools for blind children have been used in testing these children.

Modification of visual-spatial-motor tests for handicapped children

Children with neurological impairments frequently have difficulty coordinating their visual and motor systems. As a result, tasks that involve smooth eye-hand coordination are poorly accomplished. Since many self-help and school skills demand good motor coordination, it is important to include measures of visual-spatial-motor performance in a psychological assessment. The Frostig Developmental Test of Visual Perception, the Bender Visual Motor Gestalt Test, the Benton Revised Visual Retention Test, or the Sequin Form Board Figure Test are some examples of tests used for children whose motor coordination is sufficiently intact to allow them to use pencil and paper. These tests are used not only to estimate how well a child can handle paper and pencil, but also to assess the accuracy of his perceptions. Does, he, for example, perceive or "see" a square as a square, though he is unable to draw or copy one? Does he recognize the spatial relationship of one figure to another?

The Frostig is a combination of five subtests that, taken together, yield measures of visual-motor coordination, visual-perception, and spatial relations. The Benton Visual Retention Test is useful for sorting out children with defects in visual memory from those who are able to remember designs but are unable to reproduce them because of motor integration problems. The Bender Visual Motor Gestalt Test is a useful measure of perceptual-motor integration. The Syracuse Visual Figure Background Test is a modified form of the Hidden Figures Test and can be used with athetoid and spastic cerebral palsied children. Instead of requiring that the child use pencil and paper to outline figures hidden in a background of scrambled lines, the Syracuse Visual Figure Background Test projects the figures on a screen.

The fact that many handicapped children perform poorly on visual-spatial-motor tests has led some psychologists and neurologists to use such tests to diagnose the presence of "brain damage" in a child with a school or behavioral problem. There is little if any scientific evidence, particularly in the case of hyperactive children, that this is an appropriate use of such tests. It is wise to be wary of a treatment program aimed at "correcting" a problem identified solely from a child's performance on visual-spatial-motor tests. Parents should always inquire whether other tests of their child's cognitive skills were given, and if so, how the child performed. Children who do poorly on tests of visual-motor skills may do quite well on other tests of psychological function, perform well in school, and go on to lead a productive, successful life. The term "brain-damaged" in this instance is an unfortunate and unnecessary stigma. A program designed to help remedy a child's problems should be based on his performance on a battery of psychological tests as well as on the observations of the professionals who have had contact with him on a day-to-day basis. Before investing time and money in an elaborate therapeutic program, we suggest that parents seek a second opinion.

Tests of personal and social adjustment

The importance of the emotional adjustment of handicapped children with respect to their education, their social adjustment, and their vocational placement has only recently begun to be appreciated. As a result, there are few if any personality or emotional adjustment scales that have been developed specifically for use with these children. For the most part, clinicians rely on traditional personality tests to gain some insight into the personality or emotional stability of handicapped children. Personality tests are designed to measure emotions, motivations, attitudes, and intrapersonal characteristics. There are two major types of personality tests—projective and objective. A projective test is one that engages the children in a relatively unstructured task (i.e., a task that has no right or wrong answers). For example, the child might be asked to tell a story about pictures of scenes in which people or Disney-like characters are shown taking part in some interaction. Such tests permit an unlimited variety of responses in order to allow the child free use of fantasy and imagination. A child's "stories" or reactions to the pictures are studied by the psychologist, who then draws on her own brand of formal training and understanding of personality dynamics for interpretation. Projective tests can either be verbal—i.e., the child is given only a brief instruction such as "tell me what you see or tell me about this picture,"—or they may require the use of a paper and pencil. For example, the child may be asked to draw a picture of himself and his family or to draw a picture of a man, a woman, or himself. In all of these, the assumption is that the child's responses, whether drawn or spoken, will reflect something about his emotional response to himself, to his condition, and to people in his environment.

Objective personality tests are typically lists of questions or statements about likes, dislikes, feelings, etc. The child is asked to agree or disagree with a statement. For example, "You are often sad," or "I am an important member of my family," or "Storms scare me." The child's responses are evaluated by comparing them with the replies gathered from a sample of the "normal" population. This raises questions about how meaningful they are with respect to the handicapped child. Although such tasks do have clinical value in certain situations, great care must be exercised to ensure that they are used appropriately and interpreted only by persons who are well trained in their use and experienced in personality dynamics. Certainly one would expect that any decisions that would significantly alter the course of a child's life would not be made on the basis of such tests alone.

PSYCHOLOGICAL REPORT

After the psychological assessment of a child is completed, the test results, interpretations, observations, and other information should be discussed with the parents and also reported in written form to the individual who referred the child to the psychologist. The child's parents have the right to expect that this aspect of the assessment will be carried out promptly by the psychologist and that either the psychologist or the referring physician or agency will give them an equally prompt, clear, detailed account of the report. In most situations, it is the psychologist's responsibility to review the report with parents, answer their questions, and discuss with them the implications of the findings for their child. Parents also have the right to read the report and to have a copy of it if they wish.

Since the purpose of an assessment is to gather information that will be helpful in planning their child's future, it is essential that the psychological report be written so that it can be easily interpreted to parents. The language in the report should be standard English, and the material should be organized in a readable format with concrete examples to illustrate important points. Obviously, parents cannot ensure in advance that the psychologist evaluating their child will provide such a report, but they can insist that the psychologist explain any portions of the report that are unclear to them.

In order to acquaint families with psychological reports, the format they usually follow, the kinds of information they should include, and the kind of information they need not include, the balance of this section is presented in the format recommended and generally followed by well-trained consulting psychologists. The report is divided into distinct sections, each labeled with a heading that identifies the kind of information included in the section. The details of the report will be easier for parents to follow and to understand if the person discussing the report with the parents structures his account in the same manner. We suggest that parents prepare an outline using the headings presented below before they meet with the psychologist or doctor to go over their child's report. In this way, parents can ensure that they are given a complete, understandable, and useful account of the results of the assessment. Psychological reports may vary in length and complexity, depending on the child's age and the nature and severity of his problem, but for the purposes of illustration, we have included all of the subheadings we could expect to find in a report on a school-age child. We have omitted developmental history for the sake of brevity, assuming that the child's doctor already has this information.

Psychological reports: section headings

Reason for referral: This section should review the questions that brought about the assessment and the extent to which these questions can now be answered.

Observation of behavior: This should include a brief description of the child's behavior in the testing situation, the quality of his interactions with the examiner, his attitude toward the tests—for example, "John was interested, cooperative, and persistent in his efforts," etc., or "John was cooperative, but he was easily distracted and soon lost interest in a task. He needed almost constant supervision to maintain task focus," or "John lacked self-confidence; he looked to the examiner for approval and confirmation and ended each of his replies with 'Is that right? I'm probably wrong' or 'I hope that's right, is it right?'." The psychologist's observations about a child's behavior may be relevant to the reason for referral and contribute to her sense of whether the results obtained are an accurate reflection of the child's potential.

Tests administered: The report should state exactly which tests or procedures were used in the assessment. These should be identified by name. If only selected portions, rather than the entire test, were used, or if the methods or materials were modified to a degree that represents a major departure from the standard version of the test, the exact nature of the changes should be described and the reasons why the modifications were necessary should be stated.

Test results: Test scores (scores, ranges, mental ages, etc.) should be discussed with the parents, taking each test in turn, along with consideration of the consistency of the scores; i.e., do the results of different tests tend to support or contradict one another? The report should acknowledge scores obtained on tests that were significantly modified either in the manner of presentation or in test materials. Whether a numerical I.Q. score is included in this report depends on the psychologist's professional conviction. A great many psychologists prefer stating I.Q. scores in terms of ranges. If parents are given a numerical score, it is essential that they also be told the ranges in which the psychologist found their child to be functioning and the factors that contributed to this level of performance. This is part of a detailed interpretation that accompanies any test score. An I.Q. score is only one measure of a child's functioning, a measure of how he has performed on a particular test at a certain point in time. It is not a number that reflects all of a child's abilities, nor does it necessarily indicate his total potential for growth. The focus of any review of test scores should be to identify areas of strength and weakness suggested by specific test scores so that parents will be better equipped to understand the psychologist's interpretations.

Interpretation: This is the most difficult part of the report for a physician or a psychologist to explain to the parents and very often the most difficult for him to write. Jargon or terms that are not clearly definable are not necessary and should be unacceptable to the parents. Parents who are given information about their child in terms that are unintelligible to them should insist on clarification, even if it means getting back to the writer of the report for an explanation. Competent psychologists avoid the use of technical terms, jargon, or phrases that are apt to be misunderstood or that, at times, are unintelligible even to other psychologists. Parents should remember it is their child's future they are concerned with, and neither they nor anyone else should act upon information that is less than precise.

Conclusions and recommendations: This is the important part of the psychological report. Often it is the only section discussed with parents, although that should not be the case. This section should present diagnostic statements drawn from the test scores, interpretations, and observations made during the testing. To be most useful to parents and other people involved in the care and treatment of a child, these recommendations should be in the form of prescriptive statements appropriate to the particular child. Parents need to remind their doctor that if they as parents are to help plan for their child, they must be as well informed as possible. Labels that have not been defined precisely are apt to lead to misunderstanding, so parents should make sure they understand any diagnostic labels used to describe their child. If a diagnostic label such as "brain damaged" or "minimal brain dysfunction" is used to describe a child, parents should make sure they understand exactly what the label means to the psychologist and to the physician who ordered the test. They also need to understand that a label may influence a child's eligibility for a particular program and it may also influence his placement in school. Parents should ask for examples of the behaviors that led to the use of the diagnostic label and also for some indication of evidence that establishes an association between a certain label and a specific behavior. If little or no evidence exists, it would be wise for the parents to seek a second opinion.

Recommendations should include a realistic management plan for the child. The parents have a right to expect the psychologist to make specific suggestions about how their child's educational needs can best be met, considering the resources of the community. Parents should feel free to ask for specific suggestions about how to manage particular problems on a day-to-day basis. The psychologist should be familiar with a family's capacities and the needs of their child. If the psychological report has been carefully written and either the psychologist or the doctor has reviewed its contents with the parents in a thoughtful, deliberate manner, they should have a sound understanding of their child's abilities and assets as well as his disabilities and deficits. Moreover, parents should leave satisfied that they know the recommendations offered by the psychologist, the person(s) who will be able to carry out these recommendations, and how they, the parents, should proceed to set these activities in motion. Parents should be sure they understand who will monitor their child's progress and whom they can look to for help if modifications in his program are needed. Parents have a right to understand the reasons for recommendations for special educational placement, specific educational programs, or referrals to other professionals.

REFERENCES

Anastasi, A. 1976. Psychological testing. Macmillan Publishing Co., New York, p. 750.
Cronbach, L. J. 1970. Essentials of psychological testing. Harper & Row, New York, p. 650.
McReynolds, P., ed. 1968. Advances in psychological assessment, Vol. 1. Science and Behavior Books, Palo Alto, Calif., p. 336.
Talbot, N. B., J. Kagan, and L. Eisenberg. 1971. Behavioral science in pediatric medicine. W. B. Saunders Co., Philadelphia, p. 467.

6

THE PSYCHOLOGIST: COUNSELING AND BEHAVIORAL MANAGEMENT

The family of a handicapped child looks to the child's doctor for advice and guidance. Unfortunately, it appears that while some family physicians and pediatricians are willing to attend to the chronic physical problems that are often associated with a handicap, many doctors are reluctant to become involved with the equally chronic behavioral or functional problems that plague these children and their families. Handicapped children's problems are frequently complicated, time-consuming and more often than not frustrating, because solutions are not readily available. Therefore, if at some point a family physician suggests a referral to a psychologist, parents should not be disappointed in the doctor or feel that he wants to abandon them because he is not interested in their difficulties. Nor should they conclude that he believes that the child, the parents, or the family is so emotionally disturbed that special treatment is needed. It is much more likely that he has evaluated the situation realistically and concluded that a referral to a mental health professional, who has the time and the training to provide the emotional support, counseling, and guidance a family needs, is in the best interest of the child and the family. Parents are less resistant to the idea of a referral to a mental health professional if their family physician takes the time to prepare them properly. For example, parents have a right to know why the physician thinks they need to consult a psychologist, and what he hopes the family will gain from the counseling experience. Parents should be given time to consider the doctor's recommendation, talk it over with each other, and come to a decision that (one hopes) is agreeable to both. A parent who enters into a counseling situation unwillingly rarely benefits from the experience. Parents need to be assured that their physician intends to continue to be actively responsible for their child's medical management and that he will be in close contact with the psychologist they both select.

SELECTING A PSYCHOLOGIST-COUNSELOR

Most physicians who refer families for counseling and guidance have established contact with a psychologist or other mental health professional in the community who will accept his referrals, provide the family with the needed assistance, and keep the doctor informed of the family's progress. If the family physician recommends one psychologist in particular, parents should feel free to inquire into the therapist's background, training, and the type of therapy he offers. In some instances, a physician will suggest several psychologists and let the parents decide which one they prefer. Parents should speak with several therapists, explain their needs, and ask for a brief description of the type of counseling each therapist offers. It is also a good idea to ask about fees and whether the family health insurance plan will cover the costs of the therapy.

Professionals who make a practice of exchanging information about a child or family without the express consent of the child's parents violate that family's right to privacy. No medical specialist should release information other than to the referring physician without the written consent of parents. However, it is important for the child's doctor to share information about the child's medical problems and his treatment with the psychologist the family has chosen to consult. Detailed information will improve the psychologist's understanding of the child and his behavior and also permit evaluation of any sudden increase in troublesome behavior or marked change in the child's activity level that could be drug-related. If this should happen, the psychologist can discuss the situation with the physician so that modifications in drug therapy can be considered. For example, phenobarbital, a drug commonly used for the control of seizures, may produce irritable, overactive, aggressive behavior in some children. A change in medication may, under certain circumstances, significantly improve the family's ability to cope with the child.

By keeping in close contact with each other, the psychologist and the referring physician can work together to help a family develop the kinds of attitudes and skills that will be the most rewarding for all family members.

The psychologist as counselor

As a counselor, the psychologist's function is to provide the emotional support, guidance, and teaching that will enable basically competent families to cope with the stress-producing situations they will encounter at various stages in their handicapped child's development. Counseling is not a recommended treatment for serious emotional disturbances. Parents or children who are emotionally disturbed usually require intense individual therapy over a longer period of time than is customarily offered in a counseling situation.

On the other hand, if appropriate counseling is provided early, before the family's problems become chronic and resistant to change, there can be significant benefits for both the child and the family.

Emotional support for parents

Whether a child's neurological disorder is obvious at birth or is discovered some time in the first months or years of the child's life, the parents' initial reaction to the news is one of shock. Grief, denial, bewilderment, and even feelings of inadequacy about their own capacity to care for their child soon follow. Moods and feelings shift rapidly from complete disbelief, anger, and hope, to despair, fear, or apathy. Parents referred for professional counseling during this very difficult period have the right to expect that the individual they consult will be understanding, patient, and responsive, not only to their needs but to the needs of the family as a whole. But most of all, they have a right to expect that the counselor will be someone who will *listen* to them. The most frequent parental complaint is that professionals, particularly physicians, do not *listen* to their concerns or answer their questions as fully as they would like. Questions are often a parent's way of signaling a need to talk, and it is a counselor's responsibility to recognize the signals, answer the questions as completely as possible as often as necessary, and encourage the parents to talk about the feelings, worries, and fears that prompted their questions. Parents need to feel free to express their sorrow, disappointment, or even hostility without fear of being characterized as "bad" or "rejecting" parents. Most parents will talk about their emotions if they feel they can trust the counselor to be understanding and not judgmental. Parents do not want pity or dishonest optimism, nor should they be subjected to negative attitudes or to thoughtless criticisms of their efforts to cope with what appears to them a monumental problem and a threat to the life they had envisioned.

Sometimes the emotional distress of parents is so deep it prevents the development of loving, accepting feelings toward the child. A counselor is expected to help parents understand and cope with such feelings so that they will be able to show their child the affection and love all children need in order to feel they are wanted and accepted.

When a child's development is delayed, parents are often reluctant to talk to each other about things they have noticed that suggest that something is wrong with their child. Communication between the couple is thus limited, and by the time the evidence of delayed development is undeniable and diagnostic evaluation is sought, the marital relationship may be seriously strained. A counselor should be attentive to clues each parent reveals about the clarity of his or her recognition that their child has a problem. A counselor's task is to help parents learn to communicate with each other again, to share responsibilities, to comfort and support each other so that together they will be able to plan wisely for the child and the rest of their family. Given this kind of support, it is possible for a couple whose relationship has suffered to begin to communicate with each other again and to gradually develop a better, more mature and understanding relationship.

Channeling the parents' fears and concerns into constructive activities will help them avoid some of the emotionally damaging and financially costly experiences some parents of impaired children engage in. For example, there is a strong tendency on the part of some parents of handicapped children to go from doctor to doctor, searching for a more favorable diagnosis, or to involve their child in questionable treatment programs or other "therapies" of doubtful value, such as restrictive diets, manipulation, massage, or others of this kind. A responsible counselor should not only be familiar with the reputable programs offered in the community, but also with poorly documented programs and be able to advise the parents about the merits of any program. Doctor shopping can be minimized if the counselor will encourage parents to bring their questions to their doctor. Often only one special consultation visit is all that is required to reassure parents.

Because the early years of life are most important for neurologically impaired children, it is essential that the attention of the parents be focused on the kinds of practical things they can do to enhance the child's assets rather than on the child's deficit. The counselor's task is to help the parents understand that an impaired child will usually develop but at a slower rate. Parents will have a positive attitude about their child if they understand this concept and help him develop his full potential by providing him with opportunities at his developmental level rather than chronological age. Successful integration of a

handicapped child into the family and later into society depends not
so much on what is done for him as it does on the underlying attitudes of the parents and other significant people in the child's environment. With appropriate counseling, most parents will make a positive adjustment to their situation and, in so doing, promote a feeling
of self-confidence and strong self-esteem in their child.

Special counseling needs

If the child's condition is apparent at birth and the mother is scheduled to go home but the baby needs to remain in the hospital for
further medical attention, a hospital-based counselor should be alerted
to meet with the parents immediately. These families, particularly the
mother, need attention, information, and emotional support if the
family is to make a healthy adjustment to their situation. Generally
the pediatrician, the family doctor, or a hospital nurse will arrange for
a counselor to meet with the parents to help prepare them for the coming separation from their child. However, if such service is not offered
by the hospital, one of the parents or another family member should
ask the doctor what steps they need to take to get the proper assistance. Very often a community social service agency, such as the Visiting Nurses Association or Crippled Children's Services, will provide
counseling upon application of the doctor or the family. If the family
lives a considerable distance from the hospital, parents are entitled to
ask that the counselor arrange to see them when they come in to visit
their child so that costly extra traveling will be kept to a minimum.
When the baby is ready to go home, parents want to be assured that
they can depend on the psychologist for continued support and guidance.

Counseling for relatives

Grandparents and siblings of the affected child frequently experience
much the same emotional distress as the parents when they learn that
their grandchild or brother or sister is not "normal." Parents ought to
be able to rely on the psychologist to help them explain the situation
to family members and, if necessary, to include them in some counseling sessions. This will make it easier for the family to work together
to find ways to solve practical problems involved in the care and rearing of the child.

Counseling in regard to child rearing

All parents want their children to grow up to be happy, productive members of our society, and parents of handicapped children are no different. It is important for all parents who hope to achieve this goal to begin their efforts early, but it is particularly important for the parents of a handicapped child. Therefore, we have devoted the following section to this topic.

MANAGEMENT AND DISCIPLINE

The notion that a handicapped child's behavior reflects the physical, medical, or mental aspects of his condition alone ignores the fact that handicapped children, like all other children, usually behave in the ways they have been taught. *The basic principle that behavior is learned as a function of its consequences applies to all individuals, handicapped and nonhandicapped alike.* The uncertainty manifested by so many parents about the proper method of teaching discipline to their handicapped child most often comes from viewing their child as "special" or "unable to behave" like other children because of his disability, which they see as the cause of his unacceptable behavior. Unfortunately, in some instances, this notion may be fostered by professionals who use labels such as "minimal brain damage" or "dysfunction" as if demonstrable behavioral abnormalities are always associated with definite structural or chemical evidence of brain damage. Actually, this is rarely the case. Labels often communicate what is to be expected of a child. Brain damage or dysfunction, minimal or otherwise, to most parents means an injury to the child's brain that relieves the child and the parent of responsibility for the child's behavior. If parents are told their child has minimal brain damage and they understand that new brain cells do not form after an injury, they frequently feel that little or nothing can be done for the child. If inappropriate, overactive, distractible behavior is "explained" to parents as evidence of brain damage and if brain damage cannot be cured, then it is logical for parents to feel that their child "can't help" his behavior.

Parents who accept this idea and treat their child as if he is incapable of learning the ordinary rules of conduct add the weight of inappropriate, unacceptable behavior to the burden already imposed on their child by his neurological handicap. Moreover, since in our society parents, like it or not, are held legally responsible for damage caused by their children's behavior, failure to teach a child acceptable behavior can result in unnecessary financial and social problems for a family. For example, the parents of a hyperactive eight-year-old were held legally responsible for the damage he caused to their neighbor's home when he put a garden hose in the basement window, turned on the water, and flooded the basement. The fact that the child had been diagnosed as "hyperactive" and "minimally brain damaged" did not excuse these parents from their responsibility either to teach the child to respect the property of others or to supervise his activities. Nor were the parents of a fifteen-year-old epileptic boy who took the family car without permission, had a momentary blackout while driving, and caused a three-car smashup, relieved of their legal responsibility for his actions.

Parents must develop a realistic view of their child's behavior before they can begin the task of teaching him adaptive behavior patterns that will not only improve his chances for a full and productive life, but help him develop more normal relationships with his siblings, family, and other children.

Cures for most neurological disorders have yet to be discovered, but significant advances in treatment and long-term management have been achieved. Among the recent advances has been the application of the principles of learning to the modification of maladaptive behavior patterns in handicapped children. As a result of these efforts, it is now possible to eliminate many of the disturbing behavioral patterns shown by some neurologically impaired children and to teach children with more severe developmental delays useful self-help, social, and vocational skills. Behavior modification strategies have been used to teach handicapped children to speak, to walk, and to feed and dress themselves. These strategies have also been successfully applied to such problems as toilet training, tantrums, hyperactivity, destructive behavior, self-mutilation, shrieking, biting, head-banging, hitting, spitting, etc. With these methods, even seriously delayed children can be taught behaviors that will improve their lives and the lives of their families or caretakers.

BEHAVIOR MODIFICATION

Behavior modification offers a means by which parents can eliminate or reduce many troublesome behaviors that can interfere with a child's ability to function effectively. It also offers a method by which handicapped children can be taught the self-help and social skills they will need to make the most of their opportunities. Although it is possible to teach oneself to use behavior modification techniques from studying books, manuals, etc., it has been our experience that the learning process will proceed more smoothly if family members are taught the techniques and practice them under the supervision of a counselor trained in behavior modification. We have also found that a counselor's instructions, directions, and suggestions are much more likely to be followed if *both* parents are present during the counseling sessions rather than father or mother alone. Obviously this is not always possible, but every effort should be made by the counselor to bring both parents into the sessions. In situations where grandparents, siblings, or others are responsible for the child's care, they too should be taught to use these techniques so that the approach to the child is consistent. It is essential that parents understand that not all behavior can be changed at one time. Modification of a child's behavior must start gradually, and each step should be carefully explained by the counselor.

It is important first to understand the *basic* principles of behavior modification. The enumeration and explanation of these principles, it should be noted, is meant to encourage and facilitate relationships between parents and a counselor; it should not replace that relationship.

Principle no. 1

Behavior is any action that can be observed. The advantage of adopting this principle is that it focuses attention on what the child does instead of "trying to discover why he does it." For example: Johnny, who is four years old, whines and cries every time he is left alone for more than five minutes. "Why" Johnny behaves this way is not clear. Some of the reasons "why" might include: (a) he feels insecure; (b) he is bored; (c) he is the middle child; (d) he is afraid; (e) nobody loves him; (f) he does it to frustrate his mother; (g) there is no "father figure" in the home; (h) he has not learned to amuse himself, etc. These are subjective interpretations or guesses, and although any one or combination of them could be true, it is usually unprofitable to focus on such potential causes. To help Johnny develop independent behaviors, the focus must be on observed behavior.

Principle no. 2

To teach, strengthen, or maintain a desirable behavior, follow it with a reinforcer. There are three classes of reinforcers: (1) tangible reinforcers (things)—candy, toys, money, or money substitutes (poker chips) that can be used to "buy" desired objects; (2) social reinforcers—smiles, kisses, praise, hugs, approval, attention; and (3) activities—outings, sports, special treats, T.V., free time, etc. The following is an example of the use of social reinforcers: Tommy picks up a toy and puts it away. The mother *smiles* (social reinforcer) and says "Tommy, you take good care of your toys. That pleases me [social reinforcer]. Thank you" (social reinforcer—praise). Tommy is likely to put another toy away as this attention and praise strengthens the behavior. When teaching a child a new skill or when first starting to modify an inappropriate behavior, reinforce him after each correct response. For example: Ben, a six-year-old child with mild cerebral palsy, needed to learn to dress himself instead of dawdling until his mother took over and dressed him. This was faster for his mother but it did not teach Ben a necessary self-help skill. When Ben's mother decided that he had to learn to dress himself, she told him that it was time for him to learn to do more things for himself. That night, mother and Ben put out the clothes he was to wear in the morning. Next morning, mother said, "Let's see how many things you can get on while I start breakfast" and left Ben, saying "I'll be right back." When mother returned in two minutes, Ben had his pajamas off and his underpants and shirt on. Mother praised Ben, saying, "You are a really fast dresser. I am really pleased. You keep on dressing while I look after breakfast." A short time later mother returned again and again reinforced Ben's efforts and ignored any dawdling. She was sure to reinforce each additional piece of clothing Ben managed to put on. After several weeks of following this approach, Ben was dressing himself without dawdling or waiting for his mother. When this was happening consistently, Ben's mother began to simply comment on how nice Ben looked and how proud he must be of himself. *To teach a new skill, reinforce every step along the way; later, slowly start to reinforce fewer and fewer of the intermediate steps.* At this point the reinforcements should be given in an unpredictable pattern. To teach a new skill, reinforce every step each time it is done; to maintain an established behavior, reinforce the entire act intermittently.

Principle no. 3

To weaken undesirable behavior, follow it with a punisher (a negative consequence) or a period of time-out. Punishers or negative consequences include: (1) physical punishment—slaps, hits, spanking; (2) social punishers—scolding, disapproval, ignoring, frowns, criticism, ridicule; (3) tangible punishers—fines, removal of objects, toys, money, or money substitute; (4) restriction of activities—no sports, T.V., play-time, etc.

Time-out is a special technique that involves placing the child in a specifically designated spot in the house for a short period of time (five to ten minutes). This serves to isolate him from possible rein-forcers or opportunity to earn a reinforcer. Time-out is a particularly valuable technique, because unlike spanking or whipping it avoids parental modeling of aggressive behavior. In addition, because it is applied immediately after the inappropriate behavior, the child learns the cause-and-effect relationship between behavior and its conse-quences. Used properly, time-out is limited to five to ten minutes (or long enough for a child to calm himself), after which the child can rejoin the family circle where most reinforcers are provided. This avoids banishing a child to his room for an unlimited time with no clear indication of what he must do to rejoin the family. Repeated banishment of this sort carries with it the message "you are unaccept-able" rather than the specific behavior is inappropriate.

For example: Jimmy grabs his sister's toys but will not share his own. The mother learned to use time-out. Each time Jimmy grabbed Sally's toys, mother gave him one warning. She said "Stop grabbing and give the toy back." If Jimmy did not heed the warning, mother immediately took him to a chair she had placed in the utility room, turned on the light, seated him, and said "We don't grab from others and we share our toys. You sit there for five minutes and then I will let you come back out to play. I know you will be able to show me that you know how to share and that you will not grab all the toys. When the bell on the timer rings, I will come and get you." Then mother left, closing the door behind her. By being paired with the punisher, the warning will gradually act as a mild punisher and re-duce the need for stronger methods. By following this procedure con-sistently mother was soon able to teach Jimmy how to share his toys with his sister and with other friends.

Principle no. 4

Punishers are effective teaching tools only if they are used correctly. Certain rules need to be remembered if punishment is to be used ef-fectively:

1. Physical punishment should, in most cases, be reserved for be-haviors that may result in serious injury to the child or to others. For example, a "No, we don't go into the street" followed by a couple of sharp swats on the bottom can teach "street safety" fairly quickly.

Giving praise for staying out of the street will reinforce "safe behavior."

2. One warning signal, usually a short command ("Stop," "No," "Put that back") should be given before the punisher. As in the example we just described, the command gradually will act as a mild punisher and reduce the need for physical punishment or stronger methods.

3. Punishment should be given immediately if this initial warning is not heeded. Punishment is most effective if it follows immediately after the misbehavior. This prevents avoidance or escape from the consequences of one's behavior and teaches the cause-and-effect relationship of behavior and its consequences. For example: Billy, a four-year-old boy, was very difficult for his mother to manage. He was an easily frustrated, demanding, and unruly child. When Billy misbehaved, his mother would attend to him, trying to explain to him why he should not behave as he did. Sometimes she tried to placate him by offering food or a toy in the hopes of averting a tantrum. Occasionally mother tried to put Billy in a chair or in a corner as a punishment, but this usually stimulated violent tantrum behavior that worsened until mother "gave in." When Mother gave in she would say "Okay, Billy, you wait until your dad comes home. You will get a real spanking, and then you'll be sorry." When dad came home mother would tell him about Billy's behavior with the expectation that he would spank the boy. Father, who resented being cast in the role of the "mean guy" would, depending on his mood, either coax Billy into saying "I'm sorry for being a bad boy" to his mother, or argue with his wife about whose job it was to "raise" Billy. In any case a spanking was rarely delivered—which was just as well, since it would not have "worked" anyway.

Billy's objectionable behavior worsened and Billy's mother finally sought help. After a period of observation mother was instructed to do the following: when Billy misbehaved, she was to tell him to stop what he was doing (a warning signal). If Billy did not stop, mother was to remain calm, not argue or threaten, and immediately put Billy in his room and shut the door (punishment for Billy). He was told he had to stay there for five minutes (mother set a kitchen timer) and if he was quiet when the timer rang she would let him come out. If he was in the midst of a tantrum when the timer rang, he had to stay in his room until he was quiet for a short period of time before he could rejoin the family. Mother was to follow this procedure *each* time Billy exhibited certain specific behaviors (e.g., kicking or hitting family members or friends, grabbing his sister's things, throwing objects, spitting at people, teasing the dog). Mother was also taught to rein-force Billy for playing nicely. This meant that when Billy was playing nicely, mother was to go to him, give him a hug and a kiss, praise him for playing appropriately and show him attention and physical affection (reinforcement for Billy). Billy's rate of misbehavior dropped sharply and within a month Billy's mother became more confident of herself, she learned to give a warning signal and to act immediately if her warning was not heeded. The verbal signal ("Stop that") soon acted as a mild punisher because it was closely associated with being placed in time-out (punishment and no opportunity to take part in reinforcers). Praising and giving physical attention to Billy when he was behaving appropriately increased the periods of acceptable be-havior.

4. Taking away reinforcers as a punishment is effective only if there are clear-cut ways for earning the reinforcers back again. In the example above, Billy learned that if he was quiet for five minutes he could come out of time-out and have the opportunity to take part in activities and receive attention and affection from his mother.

5. Punishment should fit the misdeed. Taking away Jimmy's bike for a month because he failed to take the trash out won't teach him to be responsible about trash removal, but setting a rule that trash must be out before Jimmy has his dinner makes it easy to be consistent and gradually teaches Jimmy responsible behavior.

6. Punishment should punish the child and not the parent. For ex-ample, keeping Richard in for a week punishes mother as well as Richard if she must give up desired activities in order to supervise him at home.

7. With the possible exception of a swat on the bottom to teach safety rules, punishment should depend on taking away reinforcers or opportunities to earn reinforcers rather than on physical pain.

Principle no. 5

A punisher paired with a reinforcer will tend to strengthen undesir-able behavior and weaken desirable behaviors. Never follow undesira-ble behavior with a reinforcer. The following example illustrates this principle. Johnny, a moderately retarded child, whines and cries as soon as mother is out of his sight. Mother goes down into the base-ment to put a load of laundry into the washer. Johnny whines and cries as soon as she leaves the room. Mother runs back up stairs say-ing "What's the matter, Johnny?" She checks him, finds nothing wrong. Mother shakes her finger at Johnny and says "Now you be a good boy. Here's a cookie [reinforcer]—okay?" Mother runs back down to the basement. Johnny finishes the cookie in short order and starts to whine. Mother calls from the basement [attention reinforcer]: "I'll be finished soon. Now you stop that whining if you know what's good for you." Johnny starts to cry. Mother rushes back up. "What's the matter with you? You're just a crybaby [punisher]. Mommy will be right back." Johnny gets another cookie [reinforcer] and mother dashes back down the stairs. This same sequence may be repeated several times until mother, out of breath, angry, and resentful, loses her temper and slaps Johnny. Johnny reacts by shrieking at the top of his lungs. Mother, ready to do anything now to turn off the shrieking, decides she may as well stay upstairs with Johnny and finish the laundry after dinner when her husband will be available to stay with Johnny, although she had wanted to use that time to do a bit of shopping. Each time Johnny whined mother bounced back to his side like a ball on a rubber string. Although after several trips up and down the steps mother lost her temper, yelled at Johnny, and threatened all manner of punishment, she did, in fact, reinforce the very behavior she found undesirable. In this example it is clear that reinforcements (verbal attention, cookies, and the mother's presence), when combined with punishers (scolding, ridicule, slaps), negated their effects. Punishers, although unpleasant, were not as powerful as the positive effects of attention and other reinforcers that mother provided. This approach to Johnny's behavior served to teach Johnny not only that whining and crying pay off, but that increasing maladaptive behavior, that is, screaming and shrieking, will produce even greater reinforcement (in this case, the mother's continued presence).

Principle no. 6

Consistency is the key to getting rid of inappropriate behavior. Although we've touched on this principle before, it deserves repeated discussion. Consistency means that every time a specific misbehavior occurs, the same response will follow. The most frequent admission made by parents seeking help for a child with behavior problems is "I know I ought to be consistent, but I'm not."

Example: Johnny's mother said "Sometimes when he whines I ignore him and if I keep it up, he will eventually stop whining and start playing. At other times I feel sorry for him and I give in. If he cries or begins to shriek I feel I have to comfort him and play with him until he calms down. My husband is no help—he is too strict. He says I baby Johnny and sometimes he won't do anything with him so I feel that even if I spoil Johnny a little, at least he knows *I* love him."

The chance that Johnny will learn not to whine in the face of such inconsistency is extremely poor. Not only is mother actually rewarding the undesirable behavior (whining), she is rewarding increases in maladaptive behavior. Johnny learns to scream and shriek and produce even more objectionable behavior to get what he wants. It is also likely that he will learn to play one parent against the other to achieve his goals. Disagreement over who is "right" about Johnny will probably result in parental bickering, frustration, hostility, and in some instances, serious damage to the couple's relationship.

Principle no. 7

It is easier to teach children to be responsible if they are given clear rules to follow. A rule should apply to certain behaviors that one hopes will ultimately become habits; for example, brushing one's teeth, taking a bath, not running into the street, not playing with matches, doing homework before watching T.V., behaving courteously, using good table manners, or calling home if one is going to be late. A rule is defined as a predetermined statement regulating certain behaviors. To be effective, a rule must conform to the following criteria:

1. A rule must be fair. It is unfair to expect Richard to keep his room clean if everyone else uses it as a dumping ground.

2. A rule must be clear. It must always specify a *behavior* and a *consequence*. For example, if the rule is that Mary must *clean her room on Saturday* (behavior) before she may go out to play (consequence), mother must indicate what she means by clean. In some families, to clean a room means to pick up toys, hang up clothes or place them in the hamper, dust furniture, make bed, etc. In other families, to clean a room simply means to pick up toys. If details are not spelled out exactly, the requirements for fulfilling a rule may vary depending on mother's mood, dad's headache, etc. In this way debates with the child about what the mother meant are avoided. If a child is slow or retarded, a rule may have to be made even more simple. A clean room may initially mean all toys are picked up. Other elements such as hanging up clothes or making the bed can be added after the child has learned to pick up his toys, depending on his abilities. When first teaching a rule, it helps to add an extra reinforcer. For example, if Mary cleans her room properly, she earns five points toward taking a friend skating.

3. A rule must have a time limit. The trash must be carried out before Willy gets dinner. There will be no T.V. on school mornings *before* the child is dressed, has his bed made, breakfast eaten, and teeth brushed. This relates the desired reinforcer to some event(s) that must take place shortly beforehand.

4. A rule must be enforceable. A parent must be able to apply the consequences for breaking a rule each and every time the rule is broken. For example: "Billy can't watch T.V. for three days," says mother, who works and gets home two hours after Billy. For those two hours, the mother's rule is not enforceable, unless someone else is home to supervise the T.V. or the plug is removed from the set. Rules should be formulated in such a way that they eliminate arguments and plea bargaining.

5. It must be easy to determine whether a rule has been followed. Consequences for following a rule should be easily applied and known to be reinforcing. Consequences for not following a rule should be easily applied and known to be effective. For example, Arthur must feed the dog and put out fresh water for it before he has breakfast. Compliance is easily determined by checking the dog's bowl. Consequences are known to be effective. Breakfast is Arthur's favorite meal. The consequence is easily applied—Arthur's breakfast is withheld until the dog is fed.

Using rules

Whenever possible, parents should discuss all rules between themselves, make adjustments, and be in agreement about a rule before it is presented to the children. This approach limits impulsive, unreasonable, unenforceable rule making. When starting a behavior change plan, it is best to begin with one rule and wait until it is obeyed consistently before announcing more rules. Use the same reminder or signal when teaching a new rule and gradually fade the signal out as compliance becomes established. When a rule is broken, the child must be required to perform the correct behavior. Parents should ignore protest, not nag, and avoid negotiations. Parents should reinforce independent rule compliance promptly.

The following is a list of terms and their definitions that parents should become familiar with. Parents involved in a behavior change program with a counselor will undoubtedly learn the meaning of these terms; however, we have included them here for the benefit of all readers.

Consequence Events that follow a specific behavior that can strengthen or weaken that behavior (rewards, reinforcers, punishers).

Reinforcer Desired or "enjoyable" consequences; an event or reward known to be desired or enjoyed by the person who receives it.

Punisher Unpleasant consequence (time-out, swat on the bottom, fines, loss of privileges, etc.).

Warning signal Usually one or two words (e.g., "Stop," "No," "Hot," "Wait," "Don't run in the street"). Always use the same warning signal for the same behavior.

Log A written narrative account or report of the child's and the family's activities and interactions with one another.

Frequency count The number of times a specific behavior occurs in a given period of time.

Problem behavior Any behavior that parents or others who must interact with the child find troublesome because it disturbs everyone else, interferes with the child's ability to function, is physically dangerous to the child or others, is so unpleasant or disruptive it causes the child to be excluded from normal activities or limits the child's independence.

Target behavior A specific behavior selected to be reduced (e.g., tantrums).

Preferred behavior. A specific behavior selected to be taught (e.g., dressing or feeding oneself).

Intervention action A selected specific response to be applied every time a target behavior occurs. A detailed plan designed to reduce certain behaviors and to build preferred behaviors.

The principles of behavior modification sometimes produce uncomfortable or hostile feelings in parents or other caretakers. Many families need professional advice about what their child should be able to do at his age, given his disability. This often requires the combined help of a physician, specialists in physical and occupational therapy, and a psychologist. Generally, it is the physician who should counsel the family about the child's abilities after screening the reports of these other medical professionals. However, a physician rarely teaches parents how to use behavior modification techniques or helps a family set up a sequence of specific goals for the child to achieve.

Even if parents understand that their child is physically and mentally capable of better behaviors, or of acquiring new skills by behavioral training, they often are reluctant to use this tool. They frequently make comments such as "It's like training an animal." (It is! Human beings are animals and learn many behaviors in the same way as other species.) "I don't like to bribe my child to do good." (Why not? Isn't it better that he will *eventually* learn the benefits of appropriate behavior or skills and be able to produce them without an immediate external reward than that he never learn to do these things at all? The temporary use of a reward or reinforcer is just the deliberate use of a device by which infants and young children learn most behaviors that allow them to adapt to their families and peers.) "My child will hate me if I am always punishing him." (He certainly will! That is why parents must not continuously punish their child but instead single out *one* behavior to be changed at any one time. It is also why appropriate behaviors must be rewarded with affection. A child must comprehend through the actions of his parents that punishment is associated with a specific, well-defined behavior and not with lack of love or interest in him.)

Generally, when a family is interested enough to enroll in a behavior modification program, these fears subside very quickly.

Setting up a plan for behavior change

As we noted earlier, some people who are having mild difficulty managing a child's behavior could overcome their problem if they carefully followed the principles outlined above. However, if a child's behavior has become intolerable to his parents, teachers, and everyone else in his environment, we recommend that a counselor trained in behavior modification help the family.

Parents who seek professional help in coping with their child's behavior should expect to be asked to make themselves available for a fairly lengthy interview with the psychologist.

PARENT INTERVIEW—INFORMATION GATHERING

The purpose of the initial interview with parents is to provide information for the counselor. It is even more important, however, that this occasion be used to establish a trusting relationship between counselor and family. Before the initial meeting with parents, many counselors will ask parents to provide them with the available data about the handicapped child (e.g., medical history, developmental or psychological evaluations, contacts with service agencies, school information, comments from teachers or other professionals). They will review this information for clues to the present problem. However, as we indicated earlier, we suggest that parents meet with the counselor first to discuss their problems briefly and decide whether behavior modification is the most appropriate therapeutic approach to their problem and if the counselor is someone they can work with comfortably. Parents have a right to decide whether they want to involve themselves with a particular counselor before they give permission to release personal, financial, and social background information about their family to yet another professional. The parents can then arrange to have information from other agencies, schools, hospitals, and doctors released to the counselor. Parents sometimes feel that there is a real danger that professionals could develop a negative attitude toward a family or child from misinterpreting an earlier report, from reading someone's stereotyped conclusions about why a child is behaving in a certain way or a description of the family and the child's problems that is distorted by personal bias. Parents are correct in assuming that this could happen. If it does, it would hinder the establishment of a good working relationship between counselor and parents. Counselor and parents will be better able to work together to identify the family's concerns and to develop ways to cope with its problems if everyone starts fresh, without preconceived notions about the other. Background information from other sources can be supplied later.

Obviously the range of questions asked in any initial interview will vary, depending on the situation. These may include the age of the child; whether there are other children in the family; who, besides the parents and siblings, lives in the home; if the mother works outside the home (if so, who takes care of the child?); does the family have relatives living close by who can provide support?, etc. Once this background information has been covered, parents should expect the counselor to begin to direct attention more specifically to family interactions, how the child interacts with the rest of the family, and what they as parents are most concerned about in regard to the child.

Changing a child's behavior depends on being able to change the way significant people (parents, grandparents, caretakers, etc.) in the child's environment react toward him. Parents' responses to their children's behavior are influenced by the values they hold and their goals for themselves and their children. Consequently, both parents should talk openly with the counselor about these very personal matters to determine whether both mother and father share the same viewpoints, to examine how they teach their values to their children, and to identify the strategies they use to reach their goals. If the parents have serious disagreements about the importance of certain values, different goals for their child, or major differences about how children should be raised, these issues will have to be resolved before a plan for behavior change is developed. Since we are concerned with families who have a handicapped child, the parents, with the guidance of the counselor, will need to examine how the handicapped child fits into the family picture—what the child's role in the family is, whether he is expected to contribute in any way, and what the parents' goals for the child are. Is the handicapped child the pivotal force around which the family revolves, or, at the other extreme, is he isolated and ignored? If there are other children, do they have behavior problems? Do parents feel that the presence of a disabled child in the home is a negative influence on the behavior of the other siblings? Both parents have the right to express their particular concerns about the child, what they think they could do to change things, what they think their spouse could do to help. Sometimes one parent is reluctant to talk openly in the presence of the other parent for fear that feelings will be hurt or an open confrontation will result. For example, the father may feel that his wife devotes her energy and attention exclusively to their handicapped child and that his needs and the needs of other family members are being neglected. He may not know how to express this without sounding unappreciative or inconsiderate to his wife. A father may also feel that his wife expects less independent behavior from their child than he is capable of. The mother may perceive her husband as uninvolved with the child, too demanding, too strict, or lacking in understanding of the child's limitations. The father, in the mother's opinion, acts as if the child's handicap does not exist, and that the child would be normal if the mother did not make a "spoiled mama's baby" or "sissy" out of him. The truth may be somewhere in between the two perceptions. Therefore, each parent should be given an opportunity to talk with the counselor alone. If this is not offered, a parent should feel free to arrange to talk with the counselor alone. The counselor, after talking with both parents alone, will be better equipped to provide them, as a couple, with the guidance and support they will need to learn to work together to improve the situation. Plans for change will probably fail unless both parents are convinced that a given plan will be in their child's best interest.

After the initial interview is completed, parents should ask the counselor to summarize the concerns they have expressed and evaluate their seriousness. The counselor should be able to restate the parents' concerns, identify those he thinks are appropriate, and explain why they have reason to be concerned. If the counselor can clarify the parents' worries and put them in proper perspective, they are much more likely to follow his guidance. Occasionally, young parents of a neurologically impaired child, who are unfamiliar with the range of "normal" age- and stage-related behaviors, will mistakenly assume that because their child is impaired, any annoying behavior he exhibits is necessarily abnormal. In their desire to be "good" parents and "do the right thing," they are apt to see problems where none exist. These parents deserve counseling, but it should be brief (one or two sessions) and directed toward broadening their understanding of childhood behavior and showing them how the behaviors they have observed in their child fit into the expected developmental pattern.

Parents sometimes complain that a counselor will focus on behaviors that they consider unrelated to the problem that prompted their visit. For example, repeated parent-teacher conferences about a child's misbehavior in school will often push parents to seek professional assistance. The child may also have behavioral problems at home, but parents are willing to tolerate them or may not recognize that much of his behavior at home is the same as his school behavior. In such situations, parents will expect the child's school behavior to be given priority, and they will be annoyed and frustrated if the counselor indicates that it will be necessary to develop a unified, consistent approach to the child's unacceptable behaviors, whether they occur at home or at school. Noncompliance would, for example, have to be dealt with in the same way by parents and teachers alike. A counselor generally has little impact on a child's school behavior unless the teacher, the parents, and the counselor work together as a team. Therefore, parents should make sure that they and the counselor are in agreement about the focus of their efforts before they involve themselves in a counseling program.

If parents choose behavior management guidance, they should expect to be asked to take an active part in planning and carrying out the program. The counselor will expect parents to learn how to observe, describe, and count a specific behavior. They may also be expected to keep daily accounts (a log or diary) of the family's activities and the interactions of the family members, noting in particular the reactions of the "target" child to the events of the day. The written logs or accounts should include not only detailed descriptions of the child's behavior, but also of the events and situations that immediately preceded the child's inappropriate reactions and the consequences for him and others that followed as a result of his behavior. The accounts should also include descriptions of compliant or appropriate behaviors and the consequences of such behavior for the child. The counselor will expect each parent (or a single parent and helping caretaker) to be willing to take responsibility for keeping such a record and bringing it to the weekly meetings. Parents can anticipate that after the first information-gathering interview, the next two counseling sessions will be devoted to analyzing the logs to identify the reinforcers that are maintaining the troublesome behaviors. As the inappropriate behaviors and the maintaining reinforcers are identified, parents should understand that they will be expected to change their own reactions to the child's misbehavior so that the reinforcing aspects of their own behaviors are eliminated. For example: Alice, an eight-year-old mildly retarded child, was unwilling to take responsibility for self-care activities although she was capable of doing so. Mother had fallen into the habit of waiting on Alice (dressing, bathing, and feeding her), because she found it quicker and less troublesome than insisting that Alice take care of herself. Periodically, mother would decide that it was time for Alice to be more independent, but Alice, who was not particularly interested or motivated to comply, would dawdle while mother nagged, scolded, and threatened, until finally mother gave in. If dawdling didn't work, Alice would throw herself to the floor, shrieking and kicking, until mother gave in. Alice behaved similarly at mealtime, dawdling and playing with her food while mother coaxed, nagged, and cajoled her to finish her meal. These efforts typically resulted in a tantrum and a disrupted, unpleasant meal for the family. Alice's parents sought professional help to cope with the child's behavior.

After the initial learning period, Alice's parents and the counselor reviewed the written accounts the parents had accumulated over the two-week period. Analysis of the logs showed that inconsistent requests and reactions from the parents and other family members were, in fact, acting to reinforce the very behaviors they reported as troublesome. The family members failed to be consistent in their expectations for Alice to assume self-care responsibilities, attended to inappropriate behaviors, and were inconsistent in the application of consequences. With the counselor's guidance, the parents learned to change their behavior so that they no longer reacted inconsistently to Alice's inappropriate behavior and learned to react positively to Alice's appropriate behavior. Alice began to dress, feed, and care for herself in a matter of three weeks. She had tantrums much less often. As Alice's behavior improved, she began to show greater interest in learning and her schoolwork improved. She began to interact more often with her peers, taking part in social activities with her classmates.

Parents of children with neurological handicaps are entitled to programs and services designed to help them learn to cope with behavior problems and to teach them how to use behavior change techniques. Parents should understand that while the techniques of behavior change sound deceptively simple, they are, in fact, extremely difficult for parents to carry out day after day with the degree of consistency and commitment necessary to achieve lasting changes in behavior. And unfortunately, professionals often offer intervention strategies before they really understand the behavior and the family interactions that serve to maintain the behavior. As always, parents should be wary of any therapy, behavior modification included, that makes great claims for success but offers little in the way of supporting evidence.

Undesirable behaviors can be eliminated or modified and self-help skills taught with behavior modification techniques, but these changes cannot be accomplished without a great deal of effort on the part of the parents. The success of any program of behavior change depends on careful gathering of preliminary data, thoughtful planning, and, most important, the parents' willingness to follow the plan consistently and to support, encourage, and share the responsibility with each other wholeheartedly.

REFERENCES

Bakwin, H., and M. Bakwin. 1972. Behavior disorders in children. W.B. Saunders Co., Philadelphia, p. 714.

Becker, W. C. 1971. Parents are teachers: A child management program. Research Press, Champaign, Ill., p. 194.

Meichenbaum, D. 1977. Cognitive behavior modification: An integrative approach. Plenum Press, New York, p. 306.

Patterson, G. 1978. Families. Research Press, Champaign, Ill., p. 171.

Patterson, G., and M. E. Gullion. 1968. Living with children. Research Press, Champaign, Ill., p. 96.

Stewart, M. A., and S. W. Olds. 1973. Raising a hyperactive child. Harper & Row, New York, p. 299.

Talbot, N. B., J. Kagan, and L. Eisenberg. 1971. Behavioral science in pediatric medicine. W.B. Saunders Co., Philadelphia, p. 467.

Ullmann, L. P., and L. Krasner, eds. 1965. Case studies in behavior modification. Holt, Rinehart and Winston, New York, p. 401.

Ullmann, L. P., and L. Krasner. 1969. A psychological approach to abnormal behavior. Prentice-Hall, Englewood Cliffs, N.J., p. 776.

Wright, L. 1978. Parent power: A guide to responsible childrearing. Psychological Dimensions, New York, p. 164.

7

THE EDUCATOR

This chapter will concern itself with the issues that parents of neurologically impaired children are likely to encounter as they become involved in educational decision making. The major topics to be discussed here include: a definition of "special education" and how it applies to a handicapped child; a summary of the Education for All Handicapped Children Act (Public Law 94–142), as it relates to handicapped children and their parents; how to communicate effectively with the child's teacher; community services available to handicapped children; and working with a handicapped child at home.

WHAT IS "SPECIAL EDUCATION?"

Education that has been specifically designed to meet the requirements of children who do not learn effectively when taught by conventional methods has been given the name "special education." Individuals who have chosen to concentrate their training in this particular area are called special-education teachers. In addition to meeting the requirements necessary to become an elementary school teacher, these individuals have obtained the extra certification necessary to become a "special educator." Most often this requires advanced professional training (a master's degree) that has included supervised work with handicapped children.

The kind and degree of special education needed for a child varies with the type and severity of the neurological impairment. Some children with neurological difficulties may not need placement in a special setting but just an informed teacher who is aware of their problem and able to handle any difficulties that might arise from it. Other children, such as those with mild motor, visual, or hearing impairments, might still be able to learn in a regular classroom at a normal pace if they are provided with special materials, such as books with large type for visually impaired children, or hearing aids for hearing-impaired children. Those whose deficits are limited to specific areas will often require special help in their weak areas but otherwise be able to function normally. Some children with mild mental retardation may learn by conventional methods with modification in the rate of presentation or in the amount of repetition necessary to learn. Children with short attention spans or with behavioral disorders may disrupt a regular classroom and require a more highly structured educational setting. Children with physical disabilities may require aid in getting to and from classes. They may also have to be tested in special ways to be sure they are absorbing the information they are being taught. There are now several laws that make it mandatory for all school systems to plan individualized programs for handicapped children to ensure that they are educated to the limits of their ability.

PUBLIC LAW 94–142

Since the 1920s, there has been increasing concern about the rights of handicapped children to receive the kind of education they need. Various laws have been drafted and passed over the years to try to accomplish this goal but few have succeeded. During the 1960s Public Law 89–750, which created the Bureau of Education for the Handicapped, was drafted. From then on a series of court cases has stimulated the public consciousness toward a greater commitment to providing education for all handicapped children. Still, until recently, actual change had been quite slow.

On November 29, 1975, President Gerald Ford signed the Education for All Handicapped Children Act (Public Law 94–142) into law. This law was designed to ensure that all handicapped children receive a "free" and "appropriate" public education. It guarantees special education and any related services to meet the "unique" needs of all handicapped children between the ages of three and thirteen. In September 1979 the bill was expanded to include individuals to the age of twenty-one.

Public Law 94–142 makes it possible for local school systems to receive federal money to help them bear the cost of delivering an appropriate education to all handicapped youngsters. To receive this money, however, local school districts must comply with all the rules and regulations set forth by P.L. 94–142. If they do not, they will lose federal funding. Furthermore, local school districts must prepare detailed plans outlining how they will carry out the purposes of the law. Some of the most important provisions of P.L. 94–142 are:

Top priority must be given to handicapped children who are not presently receiving an education, and second priority to the most severely handicapped children whose education is not presently meeting their needs.

All methods used for testing and evaluation must be racially and culturally nondiscriminatory. No one test or procedure can be the only basis for making a decision about an educational program.

If children are placed in private schools by state or local educational systems in order to provide them with an appropriate education, there should be no cost to the parents.

Federal money can be withheld from a state if it is found to have failed to comply with the Act; the state can refuse to fund local school systems if they have not complied with the Act.

Thus, P.L. 94–142 states that it is the responsibility of school districts to make sure that each handicapped child receives an education that is geared to his or her own particular needs and in the "least restrictive" environment possible. Usually a regular public school classroom would be considered the least restrictive environment, but such placement is not always realistically possible for handicapped children.

One of the placements described below must be made available to every handicapped child. Parents should remember that the least restrictive environment appropriate for their own child is the situation that best meets the child's educational needs.

1. *Regular classroom:* Neurologically impaired children whose handicaps do not interfere with their ability to learn at a normal pace may very well benefit from placement in a regular classroom. Teachers in a regular classroom generally do not have special-education training.

2. *Regular classroom with a consultant:* On this level, the classroom teacher receives special assistance from an individual who has been trained to work with handicapped youngsters. The consultant's function is to help the teacher alter the educational environment within the regular classroom so that the handicapped child can function effectively and still remain in that class. For example, a child with a hear-

ing impairment might be proficient enough in language and speech to be able to attend a regular class. The consultant might meet with the child's teacher two or three times a week to make suggestions about better ways to teach the child (e.g., face the child when giving instructions) and to be available to answer any questions the teacher might have.

3. *Regular classroom with service from an itinerant special teacher:* The itinerant teacher, who may be based at a central office, travels to an assigned school and there provides the services necessary. The itinerant teacher may be a specialist in reading, speech, language disabilities, or some other field. These teachers meet directly with the child several times during the week. The child spends most of his time in the regular classroom.

4. *Regular classroom with service provided from a resource room:* The child spends a part of every day in a resource room with a specially trained teacher and the rest of the day in a regular classroom. Responsibility for the child is divided between the regular classroom teacher and the special-education teacher. Reading, spelling, or arithmetic could be taught to a child diagnosed as learning disabled in the resource room as needed.

5. *Regular classroom plus part-time special class:* This arrangement is commonly prescribed for more severely learning-handicapped children. Thus, a mildly mentally retarded child might do all of his academic work in a special nongraded classroom but go to gym, art, music, manual arts, or home economics with children his own age who are of normal intelligence.

6. *Full-time special class:* Such a placement is often the most appropriate for children whose handicap is severe; it still affords some degree of contact with nonhandicapped children, since the classroom, while self-contained, is within a "regular" school building. This arrangement allows the child to have the services of a full-time specially trained teacher.

7. *Full-time special day school:* Very often children who are deaf, blind, or who have serious behavior problems are placed in special schools where the total environment is oriented to their needs. These children are still able to live at home but need a more structured school environment. While contact with nonhandicapped peers is limited, the personnel who work with the students are highly trained. Frequently there are extra services available to the students, such as those offered by occupational therapists, physical therapists, and psychologists.

8. *Residential schools:* Certain children, usually those who are se-

verely retarded, require a total environment designed to meet their needs. Because of their physical limitations, it would be quite difficult for them to receive the training they need while living at home. Occasionally handicapped children attend residential schools because less restrictive facilities close to their home are not available.

9. *Hospital or home-bound teachers:* Occasionally children with severe handicaps need to be confined to hospitals or to their homes for long periods of time. When this is the case, it is possible to obtain the services of special itinerant teachers who will come to the hospital or to the child's home to provide the necessary instruction. In some cases, it is possible to set up two-way communication between the child's home or hospital room and the school he would normally attend in order to allow the child to continue to communicate with classmates and teacher. Parents should be aware, however, that this means of providing education to handicapped children should only be used as a *last* resort, since a child also has emotional and social needs that are best met by attending school with other children.

When considering this hierarchy of services, parents and others concerned should remember that a child should be placed only as far from the regular classroom as is necessary to secure the best educational program for his needs and then returned to the least restrictive environment as soon as possible. Inherent in this philosophy is the need for continued reevaluation of a child's progress so that a child does not necessarily have to remain in one particular type of placement for his whole school career.

The concept of "mainstreaming" or returning handicapped children to the regular classroom is reflected by the "least restrictive environment" clause of P.L. 94–142. Many parents are concerned about the "stigma" they feel is attached to special education, and they are anxious to have their children removed from these classrooms and put into regular classes. Yet, in reality, this is not always the most beneficial placement for every handicapped child. Special classes generally offer more individual attention and specially trained teachers. Mainstreaming handicapped children into regular classrooms is certainly a sound choice for many students, but the type and degree of handicap involved, plus the personality of the individual child, always need to be considered carefully. One child might feel insecure and lost in a regular classroom; another might feel challenged.

It is crucial for parents to have a clear understanding of what their rights and their children's rights actually are under this law. The term "least restrictive environment" simply means the environment closest to a regular classroom that is at the same time most appropriate for a child's needs. The intent of the law is that the system should try to fit the child, and not the reverse. The term "mainstreaming" means placing the handicapped child in the environment least restrictive for *him*. The purpose is to bring him into contact with nonhandicapped children.

Parents should take the time to familiarize themselves with the process involved in obtaining the least restrictive environment for their child, as outlined in P.L. 94–142 (see Chapter 12). An excellent source of information is the newsletter *Closer Look* published by the Parents Campaign for Handicapped Children. Parents can write to this organization for materials that will provide them with updated information about their rights and the rights of handicapped children, as well as other details about services, legislation, and programs for handicapped children.

The procedure outlined for getting a handicapped child into the best educational placement for him is as follows:

1. *Identification:* Before a child can receive any type of special education, he must be identified by his local school district as handicapped. If a child is not receiving special services and the parents feel he should be, if a child is just starting school, or if a child is being enrolled in a new school district, parents should notify the school principal of their child's handicap so that the school can arrange for an evaluation.

Occasionally, parents will not be aware that their child is handicapped; a teacher may be the first person to recognize that a child has a problem. She might notice that the child is slower to learn new information than his peers, has trouble with visual tasks, does not seem to hear directions, or exhibits some form of behavior that suggests he may require special help. When a school suggests that a child is a possible candidate for special services, it must notify the parents in writing of its desire to have the child evaluated, explain why, and ask for written consent. P.L. 94–142 also states that parents have the right to "inspect and review" their child's educational records.

2. *Evaluation:* After a child has been identified as possibly needing special education, the next step is for him to be evaluated. Most local school districts have their own evaluation units; however, some may contract the evaluation to another agency. The evaluation is provided at no cost to the parents.

If parents disagree with the results of the evaluation provided by the public agency, they have the right to request an independent evaluation. The school district must provide the child's parents or his legal guardian with information concerning where such an evaluation can be obtained. However, if the school district feels its own evaluation is appropriate, it can initiate a hearing to prove that this is the case. If the final decision is in favor of the district, parents can still obtain a second opinion; however, they have to pay for it.

3. *Development of the individual educational program:* After the evaluation is completed, the next step is to develop an Individual Educational Program (IEP) for the child. P.L. 94–142 explicitly states that an IEP must be written for *every* child who is receiving special education. After the local school district receives the information from a child's evaluation, it is responsible for scheduling a meeting for the purpose of developing an IEP for the child. Parents must be notified, in writing, of the date, time, and place of the meeting early enough to have the opportunity to attend. The notification must also include: who will be in attendance (usually the principal or his representative, the child's present teacher or a special education teacher, a member of the team who evaluated the child or some other person who is well qualified to interpret test results, the child when it is appropriate, and other individuals whom either the parents or the school wish to include).

It is *very* important for parents to attend their child's IEP meeting, particularly the first one (IEPs must be reviewed at least once a year and parents or school staff can ask for a review of an IEP at any time during the school year). Basically, the IEP is a summary of the evaluation information and serves to point out a child's particular problems, what the goals for his education should be, how these goals will be reached, and how achievement of the goals will be measured. Parents are important sources of valuable information about their child's needs, and they should be active participants in the IEP planning session. The school staff will have developed a list of both short- and long-range objectives for the child that they will present to the parents for consideration. Parents should feel free to ask questions and offer suggestions about these objectives, keeping in mind that the primary goal is to achieve the best education possible for the child.

Although IEP formats may vary from school to school, the basic components as mandated by P.L. 94–142 are:

1. a statement of the student's present level of educational performance;

2. annual goals for the student, those objectives he is expected to reach by the end of the school year;

3. short-term goals, broken down into instructional objectives that lead to the attainment of the year's end goals (suggested questions for parents: what happens if Jimmy fails to meet his first short-term

goal—will the teacher change the method of teaching? Why are some of the goals things the child already does at home?);

4. an explanation of the specific educational services to be provided;

5. information concerning how much time the student will participate in the regular education program (e.g., attending regular classroom but receiving reading instruction in a resource room);

6. expected dates for starting services and an estimation of how long they will be continued;

7. a statement describing how the child's progress will be measured at the end of the school year.

A good IEP is one that communicates in clear and simple language; the information presented should be factual and not open to different interpretations. It should include a provision for keeping parents informed about the child's progress during the school year. Figure 7.1 is a sample of an IEP that is clearly organized, detailed, informative, and useful to teachers and parents. Figure 7.2 is a sample of a poorly written IEP. Annual goals are not stated, short-term goals are vague, and it lacks explanation for instructional objectives.

Parents must receive written notification before their child is placed in a special-education program, and they must give the school written permission to do so. This notification is usually mailed to them after the initial IEP meeting. If they have participated in the planning and fully understand the reasons behind the placement, most parents feel comfortable about the decision. However, parents should be sure to visit the specific program suggested in the IEP before they accept that document. Programs and facilities vary remarkably within and between school districts, yet they may carry the same labels. For example, in some school districts learning disabilities classes contain children with serious behavior problems and/or moderate retardation rather than just children of normal intelligence who are deficient in specific school skills. Visiting the program will permit parents to judge whether it is appropriate for their child.

While some school districts automatically mail a copy of the IEP with the placement notification, others do so only upon request. Parents should request a copy of their child's IEP so that they can refer to it during the school year. School districts must also provide parents with instructions "written in language understandable to the general public" detailing the exact procedures to follow if they do not agree with the final form of their child's IEP or with the recommended placement. Although each state has different ways of handling this procedure, all must comply with the federal due process regulations set up

INDIVIDUAL EDUCATION PROGRAM

Student's Name Steve Smith   Birthdate 7-14-69   Date of I.E.P. Conference 6-2-80

Present Educational Placement Learning Disabilities Class

Team Participants:

Mr. and Mrs. Smith - Parents   Dr. White   - Psychologist
Mrs. Jackson   - L. D. Classroom Teacher   Miss Lane   - Speech Teacher
Mr. Jones   - Counselor
Mr. Heart   - Assistant Supervisor-Special Education

PRESENT LEVEL OF EDUCATIONAL PERFORMANCE:

Academic Achievement: Math   1.9 (PIAT)   Spelling   1.3 (PIAT)
                       Word Recognition   1.3 (PIAT)   Word Attack   1.0 (Woodcock)
                       Reading Comprehension 1.1 (Woodcock)

Social Skills: Difficulty interacting with peers.
                       Creates verbal and nonverbal distractions in the classroom.
                       Overly dependent on teacher for support in starting and doing tasks.

Prevocational Skills   Has trouble following directions (both written and verbal).
                       Short attention span.
                       Only completes 40% (or less) of daily written assignments.

Self-Help Skills   Needs excessive teacher attention and direction.
                       Only works independently for 5 minutes at a time.

Other: Misuses verb tenses when engaged in spontaneous speech.

ANNUAL GOALS (According to Priority)
1. Steve will increase his attention span and his ability to complete tasks.
2. There will be a decrease in the number of verbal and nonverbal distractions created by Steve.
3. Steve will increase his reading sight vocabulary.
4. Steve will increase his mathematics computational ability.
5. Steve will be able to increase his correct usage of verb tenses in his spontaneous speech.

SHORT TERM OBJECTIVES (Correspond with Annual Goals)

1. Steve will demonstrate attention to classroom activities and respond appropriately; i.e., 90% of the time he will complete assigned tasks.
2. Steve will be able to control instigation of verbal/nonverbal classroom distractions so that no more than 1 occurs per 5 minute work session.
3. Steve will both improve his word attack skills and increase his sight word vocabulary to at least a 1.8 level, as measured by the Woodcock Reading Mastery Test (Form B).
4. Steve will demonstrate the ability to compute single digit addition and subtraction problems to a 2.3 grade level on the Key Math Operations.
5. Steve's spontaneous speech will contain fewer than 10% misusage of verb tenses, as measured by taped samples of recorded conversational speech.

OBJECTIVE	SERVICES NEEDED	RESOURCES NEEDED	DATES Begin	End	Review	OBJECTIVE MEASUREMENT CRITERIA
1. Attention span & task completion	L.D. Specialist	Behavior Modification and charts.	9/80	6/81	6/81	Frequency count of attending behavior will show 25% improvement; 50% completion of written work.
2. Verbal/nonverbal classroom disruptions	L.D. Specialist	Behavior Modification and charts.	9/80	6/81	6/81	Frequency count of disruptive behavior will show no more than 1 disruption per 5 minute work session.
3. Word attack and sight vocabulary	L. D. Specialist	Language Master Dolch List Language Experience	9/80	6/81	6/81	Performance on Woodcock word identification and word attack subtests to a 1.6 level.
4. Arithmetic Computation	L.D. Specialist	Math games Cuissenaire rods Flash Cards	9/80	6/81	6/81	Performance on Key Math operations to a 2.3 grade level.
5. Verb Tense	Speech teacher	Auditory and Visual Prompts	9/80	6/81	6/81	Taped samples of conversational speech will show fewer than 10% Misuse of verb tenses.

COMMENTS: Mr. Jones (counselor) will contact parents midyear to set up a conference regarding Steve's progress.

PARENT INVOLVEMENT: I have participated in the preparation of my child's individual education plan and:

___√___ I am in agreement with it.   (or)   _____ I disagree with it.

I understand that this is an educational plan and not a binding agreement.   Signature: Mrs. Elizabeth Smith Mr. John Smith

Date: 6/2/80

Figure 7.1. This is an example of an IEP prepared for an eleven-year-old boy who was diagnosed as learning disabled. It is important to remember that these are individual plans; thus, the IEP for one learning-disabled child will most likely differ markedly from that of another. It should also be pointed out that different school districts may use different IEP formats.

INDIVIDUALIZED EDUCATIONAL PROGRAM

RE: John Doe
Birthdate: 6-30-69   Grade 5th
Address: _____ Zip _____
Phone:
LEA:
Parent(s), Guardian(s) J. Doe   M. Doe
Teacher(s) Mrs. Smith

Date of Last Vision Screening 9-20-77
(Passed, Failed)
Date of Last Hearing Screening 9-20-77
(Passed, Failed)
Attendance Record: From _____ to _____
Days Present _____ Days Absent _____

Least Restrictive Education Environment:
Regular Class - L.D. Resource Room
Projected Date of Placement:
9-2-79
Person(s) Responsible for
Development of Short-Term IEP:

CONFERENCE(S):
Chairman _____ Mr. Brown
Initial IEP Date 4-19-79
Revised IEP Date _____
Participants:
Mr. Brown
Mrs. Smith - Teacher
R. Green - Counselor
O. Rose - L. D. Reading Teacher

Code for type of Evaluation: (1)Observation;
(2) Informal or teacher made tests; (3) Formal
testing; (4) Psychologist testing

PRESENT LEVELS OF EDUCATION PERFORMANCE

Skill Area	Informal Assessment			Formal Assessment		
	Date	Type	Results	Date	Type	Results
Reading	3-79	2	2.5	4-79	3	3.0
Word Recog.	3-79	2	3.0	4-79	3	3.3
Comprehens	3-79	2	2.7	4-79	3	2.6
Composite Reading				4-79	3	3.0
WISC-R				4-79	4	105

Secondary Handicap(s): Impulsive,
poor attention

Special Health Data: _____

Other Pertinent Information: _____
Poor peer relationships

PARTICIPATION IN
REGULAR EDUCATION:

Program	Duration
Regular Class	9-79 to 6-80

PRIORITIZED LONG TERM GOALS

Improve Reading Skills

Improve Reading Comprehension

Improve interest in reading

SUPPORTIVE SERVICES RECOMMENDED

Type	Facilitator	Duration
L.D. Class	Mrs. Rose	9-79 to 12-79

Facilitator	Type of Evaluation
Mrs. Rose	3
Mrs. Rose	3
Mrs. Rose	1

CC: Parent, Teacher, Master File

_____
(Parent Signature)

Figure 7.2. This is an inadequate and incomplete IEP. The parents are not listed as being present at the conference, nor is there any evidence that the parents read or signed the IEP. There are no short term goals, and there is no indication of how long term goals are to be achieved.

in P.L. 94–142. Thus, parents are allowed a special hearing to which they are permitted to bring legal counsel in addition to any individual having "special knowledge or training with respect to the problems of handicapped children." If the hearing rules in the parents' favor, the contested changes in the IEP must be made. If the ruling is not in the parents' favor, they have a choice of either accepting the decision or removing their child from the public school system and finding private sources to educate him at their own expense.

Parents need to remember that P.L. 94–142 is still a relatively new law and that all states do not as yet have all services available. Rural areas in particular are often not equipped to offer as wide a variety of services as urban areas. A handicapped child is entitled to all the rights outlined in P.L. 94–142, but, in some instances, parents need to recognize that although efforts made by their local school district are not in strict compliance with the law, they are offered in good faith and are the best possible under the circumstances. At the same time, they must judge if the school district is actually trying to meet a child's needs or if it is being lax. Occasionally, litigation is necessary. (If parents do become involved in a due-process hearing, their child must remain in his present educational placement. If the hearing has to do with placing the child in a public school, the child is entitled to such a placement until a final decision is reached.)

4. *Placement in the least restrictive environment:* After the IEP has been developed and agreed upon, the next step is for the school district to begin implementation of the program. If the local school district is not able to provide the needed services, it must either bring in the necessary personnel or place the child in a private school or facility that is equipped to provide the services outlined in the IEP. This must be done at *no* cost to the parents. Even if the child is placed with a private agency, it is still the local school district's responsibility to make sure that the parents are notified of any IEP meetings so that they can attend, and also that one of their own representatives is present at the meeting.

As previously mentioned, the primary purpose of the IEP is to set up *realistic* educational or vocational goals for the child. Once these have been set up and agreed on by both teacher and parent, it is important for parents to trust the judgment of the teacher.

Specific educational activities and techniques vary widely, depending on the nature and degree of the child's handicap. The most important responsibility of a special-education teacher is to find and capitalize on a handicapped child's assets. Thus, a child who is able to express himself verbally though he cannot manage pencil and paper should be permitted to take oral examinations rather than written, and he should be actively encouraged to make special verbal reports on assigned topics to the class. Such activities could earn him recognition from his classmates and his teacher, which in turn would enhance his self-esteem.

EFFECTIVE COMMUNICATION WITH THE CHILD'S TEACHERS

The sooner a positive relationship is established between a handicapped child's parents and the child's teacher(s), the better it is for the child. Neither role, that of teacher or parent, is an easy one, and a good relationship between both parties encourages more effective teaching as well as parenting. But the question remains: how do parents go about establishing a good relationship with their child's teacher?

First, parents should make teachers aware of their interest, support, and willingness to discuss any matters that concern their children. At the same time, parents should strive to let teachers know they have confidence in their abilities and judgments concerning classroom matters.

Parents should also ask their child's teacher to let them know of his special accomplishments so that they and the family can share in them. Too often, busy teachers forget to notify parents of the child's achievements, and notes and phone calls home always mean "trouble" of some sort. One parent made it a practice to leave a number of stamped addressed envelopes and note cards with the teacher at the beginning of each school quarter. The teacher was asked to use these to drop the parents a few lines about special efforts or accomplishments. This small gesture indicated the parents' recognition that the teacher was busy and their willingness to make it as easy as possible for her to communicate with them.

Parents also have a responsibility to provide teachers with certain types of pertinent information. They must be sure that the teacher receives information about medical aspects of their child's handicap. Although such information is usually discussed during the IEP conference, it is wise to review it with the child's teacher before the beginning of each school semester. The child's physician is the person to help parents decide what is relevant information for the teacher. The teacher should know about the child's activities at home and about important accomplishments. For example, notifying a teacher that a physically disabled child has just mastered the difficult task of dressing himself would allow the teacher to reinforce the activity through both acknowledgment and praise. At the same time, it might give her a valuable insight into the child's abilities at home as compared with school. If parents appreciate a teacher's efforts, they should tell the teacher so. They might even consider inviting their child's teacher to share a meal with their family. This rapidly vanishing, but once popular, custom is still a good way for parents and teachers to quickly get to know each other better as individuals. Regardless of the means parents choose, if they feel a teacher is doing a good job, they should let her know! On the other hand, if parents feel that the teacher is not doing a good job, they should discuss their concerns with her and with her superiors if necessary. If this does not bring satisfaction, parents must consider the other methods of appeal open to them under P.L. 94–142.

SUPPORT RESOURCES

Parents should join a consumer organization composed of parents of handicapped children. Experienced parents can help others who are just starting out. Parent organizations offer workshops, patient training sessions, problem clinics, etc. Conferences that include parents and educators, members of state and local chapters of parents or advocacy organizations, such as the Association for Children with Learning Disabilities, the Association for Retarded Citizens, and United Cerebral Palsy, can help inexperienced parents find the support and information they need to secure the right educational setting for their child. In addition, these organizations publish newsletters and distribute other materials that contain up-to-date information about new state and federal regulations governing the rights of handicapped individuals.

Families should be alert to the benefits of early intervention. The first years of a child's life are crucial in determining what kind of an adjustment he will make to his handicap. Many communities offer infant stimulation programs. Classes exist to train parents in valuable techniques for working with their handicapped babies at home. Information about these classes can be obtained from the pediatrician or from the local chapter of a parents' organization.

When investigating schools and programs for their child, parents should also inquire about summer programs and camps. Good day camps and even overnight camps for handicapped children are becoming more common. Parents should check with their local Board of Education and the local chapter of a parents' organization for specific information.

Entering school can be a difficult time for a handicapped child. Sometimes working with a school counselor or school psychologist can help a child cope with his problems. Occasionally, local chapters of organizations concerned with a specific type of handicap provide counselors to help children with school adjustment problems. However, if the problem is severe, it should be brought to the attention of the child's physician; he can then refer the family to the appropriate professional for further care.

Recent regulation (Vocational Education Amendment of 1976, P.L. 94–482) has greatly expanded the responsibility of states to provide a vocational education for handicapped children. Special funds may now go to school districts for secondary school programs, to advanced training in adult vocational educational programs and to community colleges. A detailed discussion of this important legislation is included in Chapter 12.

THE ROLE OF THE PARENT AT HOME

Many parents, whether their child has a handicap or not, are concerned about the nature of their role in helping their child master academic tasks. Should they help with homework? This is a difficult question, since the personality of the child and those of his parents are important to the success of this type of effort. Some children do quite well when their parents give them a helping hand with a difficult assignment. In other families, the situation rapidly becomes explosive and either the child or the parent becomes too emotional and often the session ends in frustration for everyone. Some children need extra tutoring to master certain concepts or skills. Whether a parent takes on the role of "teacher" at home should depend on the parent's ability to remain unemotional and objective. If the parent finds that these help sessions end in anger or emotional upset, it would be much better to find a tutor to work with the child. (Parents might want to enlist the aid of a willing high school student.) The classroom teacher can be helpful to parents by explaining the nature of the child's difficulty and offering concrete suggestions about the kind of assistance parents can provide. If a parent finds that it is possible to work successfully with a child, close communication with the teacher will help ensure that the work done at home complements what is being done in the classroom. All at-home sessions with a child involving schoolwork should be kept short and pleasant. The sessions should end on a positive note. Involving the child in setting goals for these sessions is also helpful in building a good attitude.

Reading to children is important. Reading to a young child helps teach him that there is a difference between spoken language and written language, and also demonstrates that reading is a valued and *meaningful* activity. (Frank Smith, in his book *Understanding Reading*, stresses these two concepts as being crucial for children to understand before they become good readers.) Older children should be encouraged to read *something* daily, even if only for fifteen to twenty minutes a night. There is *no* better way to become a good reader than by reading frequently. (If children see their parents reading every night, they too become interested in the printed word. Sitting in another room watching television delivers an altogether different message.) Blind children can read Braille or listen to "talking books." Deaf children, if very young, can practice their lip reading and enjoy the pictures while being read to; older children can read to themselves.

Intellectual stimulation is not limited to the classroom. Stamp collecting, rock polishing, coin collecting, sewing, baking, and raising gerbils are but a few of the many activities handicapped children can enjoy. The activity must be interesting and fun for the child and be something at which he can be reasonably successful. Encouraging a hobby is simply another way to help build a child's self-confidence and give him something to share with others. The child should be included in trips to special events or places if it is possible and appropriate for him to attend. This does not mean that parents should not take time for themselves, but that exposure to such activities, when possible, can be richly rewarding to all children.

Finally, a note of caution to parents: Make sure you communicate to your handicapped child that you value him for himself, above and beyond his academic accomplishments. It is typical for *any* parent to praise his child by saying "You got a 95 percent on your spelling—why, Johnny, you are wonderful!" Many children make the association that the only way to be "wonderful" or a "good boy" is to do well in school. Thus, a more constructive statement would be "You got a 95 percent on your spelling test—why, Johnny, that is very good work."

BEHAVIORAL MANAGEMENT

In addition to being concerned with the type of special education that is most appropriate for handicapped children, special-education teachers must also be well acquainted with a variety of behavioral management techniques. Any group of children, handicapped or not, will have among them children who have difficulties with self-control. One or two such students can disrupt a whole class and make it difficult for others to learn. A good teacher tries to ward off potential behavior problems by using some of the following techniques:

Making sure that the child is capable of doing the assigned work;

Always having on hand extra activities that a child can do if he finishes the assigned work early;

Keeping assignments to a length that ensures that individual students will be able to keep their attention focused on them;

Whenever possible, including numerous activities that focus on an individual child's strengths in order to help build self-confidence and positive self-feelings;

Trying to include the child, when possible, in setting realistic goals for himself.

The general principles of behavioral management are discussed in Chapter 6.

REFERENCES

Bryan, T., and J. Bryan. 1978. Understanding learning disabilities. Alfred Publishing, Sherman Oaks, Calif., p. 382.
Buscaglia, L., ed. 1975. The disabled and their parents: A counseling challenge. Charles B. Slack, Thorofare, N.J., p. 393.
Dodson, F. 1970. How to parent. Nash Publishing, Los Angeles, p. 444.
Gearhart, B. R., and M. W. Weishahn. 1980. The handicapped student in the regular classroom. C. V. Mosby, St. Louis, p. 303.
Shea, T. M. 1978. Teaching children and youth with behavior disorders. C. V. Mosby, St. Louis, p. 343.
Smith, F. 1978. Understanding reading. Holt, Rinehart & Winston, New York, p. 260.

To write for:

From: *Closer Look*, Box 1492, Washington, D. C. 20013
Common Sense from *Closer Look* (A publication of the Parents' Campaign for Handicapped Children and Youth).
Know Your Rights—and Use Them!
P.L. 94–142: What Does It Mean?

From: Wisconsin Vocational Studies Center, University of Wisconsin—Madison, 964 Educational Sciences Bldg., 1025 West Johnson Street, Madison, Wisconsin 53706

    Tindall, L. W. 1978. Vocational education resource materials, a bibliography of materials for handicapped and special education, 3rd ed. (a comprehensive collection of resources).

8

THE PHYSICAL
AND OCCUPATIONAL
THERAPIST

Many parents feel overwhelmed and helpless at the prospect of dealing with their child's handicapping condition day in and day out. Everyday tasks can become enormously time-consuming. For example, if an infant's neurological problems are reflected in motor difficulties, he may be a poor feeder. When he does not progress in his feeding skills quickly or easily, prolonged feeding times and parental frustration often result. If an infant or young child is easily upset, or the opposite, difficult to arouse, then parent-child interactions such as picking up the baby and cuddling, bathing, and playing with him can become difficult and potentially unpleasant exchanges. Dealing with people outside the immediate family can be stressful as they ask questions, express opinions, or stare at a handicapped child.

When a child's neurological problems are reflected in delayed development, impaired ability to move, and/or poor interaction with his surroundings, the family physician or specialist may refer the child to a physical or occupational therapist. Pediatric physical and occupational therapists are specialists in motor development. Their responsibility is primarily one of educating parents. They assist parents in helping their infant or child achieve his fullest potential in motor skills and in preventing secondary problems, such as limb or trunk deformities or joint stiffness.

REFERRAL TO PHYSICAL OR OCCUPATIONAL THERAPY SERVICES

Most states require that a physician make a written referral for a child to be able to receive physical therapy or occupational therapy services.

A referral should include information on the child's age, diagnosis, special medical precautions, and a general description of the treatment desired and goals for the therapy program. Children are usually referred for physical or occupational therapy by a family physician, pediatrician, pediatric neurologist, or orthopedic surgeon after a thorough diagnostic examination.

In many facilities, physical and occupational therapists work as part of a team and evaluate the child during the initial diagnostic work-up period. The health team of physician, psychologist, physical or occupational therapist, nurse, social worker, and speech therapist meet following the diagnostic testing to present their findings and plan for the child's care. In this type of facility, the child may be admitted to the hospital or may come in as an outpatient for a week of diagnostic tests. The physician or the team then meets with the parents to discuss their results, recommendations, and plan of care. When a child is referred for physical or occupational therapy outside a facility such as the above, parents should be able to provide the therapist with information about their child. This should include a summary from the physician and other professionals, a copy of the child's medical record, and any psychological test reports. Parents should expect a therapist to take a social and family history if one is not available, as this information is considered in formulating a home-based treatment plan.

INDICATIONS FOR TREATMENT

Treatment is usually indicated for one or more of four major reasons: to help the child achieve his maximal developmental potential (habilitation); to prevent deformities (prevention) or maintain existing skills (maintenance); to help the child regain use of an extremity, improve strength, posture, or function that the child once had achieved and has lost (rehabilitation); or, if the child has reached a plateau in his ability to progress in motor skills or function, to teach compensatory skills (remediation). Therapy can be beneficial at each of these levels; however, many therapists feel that a child has most potential when therapy is started as soon as the problem is identified and they have the opportunity to work with the child from the aspect of prevention, habilitation, or maintenance rather than later for rehabilitation or remediation. When therapy is preventive it may be regarded as enrichment, and when it is remedial it is regarded as corrective.

Some reasons for referring an infant or child might be to prevent the development of deformities that can be caused by limb stiffness due to spasticity; to help a child control his head movements; to improve his balance to the point that he can sit, crawl or stand; to teach him to roll and change positions; to improve his feeding skills, such as chewing or the coordination of swallowing and breathing; to improve the use of his hands; to help him learn dressing skills or to provide special equipment and training in its use in order to enlarge the scope of a child's activities.

LOCATING A THERAPIST

A child's neurological handicap and motor disabilities may require the services of physical and occupational therapists with specialized training in pediatrics. Therapists who work primarily with adults may not have the educational background, the experience, or the ability to work with a child with a developmental disability. As a result, parents sometimes are told: "Bring your child back when she can walk," or "There is nothing more that I can do for your child," or "I do not know how to help you with a handicapped child's therapy program." Even if a therapist is unable to treat a child, she is still obligated to assist a family in locating a facility where their child can receive treatment, or to arrange a consultation with a physical or occupational therapist who specializes in pediatric problems. The family may need to travel to the closest children's hospital to obtain an evaluation and instructions for a home program. If possible, the follow-up care should be arranged locally. Once the child's program is established, the therapist at the referral center can act as a consultant to the local therapists.

Other sources of help include regional state diagnostic pediatric centers, children's hospitals, the Easter Seal Society, United Cerebral Palsy, the United Way or other agencies who provide services to the handicapped, or state physical and occupational therapy associations. The therapist who has specialized in pediatrics either has done graduate study beyond the four-year college program in physical or occupational therapy or has worked under a pediatric therapist for a minimum of two years.

PROGRAMS FOR CHILDREN WITH PHYSICAL DISABILITIES

Therapy can be offered through a center-based program or a home-based program. Each program has its merits. Perhaps the most comprehensive approach to the management of the child's disabilities and his family's needs can be achieved through a combination of both.

A center-based program is one that is offered through a hospital, rehabilitation center, day-care center, nursery school, or similar facility where medical services are provided. The center-based program involves professionals such as physical and occupational therapists, psychologists, nurses, special educators, social workers, and speech therapists. They often have trained paraprofessionals working with them. For example, a physical therapist might evaluate and plan a therapy program for a child but may supervise a physical therapist assistant or aide in the actual treatment of the child. Parental participation varies in different center-based programs. Mother-infant group programs in which the mother works only with her own child are a common type of center-based program. Other programs for infants train the parent as a therapist's or teacher's aide; the parent may work with her own child and with the other children depending on the setting. Some center-based programs offer parental observation of therapy without direct contact with the child. In these sessions staff members can aid parents by explaining pertinent aspects of the treatment program and the child's behavior as they occur.

Group discussions or meetings for parents are offered by many center-based programs. Some groups discuss parental feelings and attitudes; others deal with child management or future educational placement for the child. Some of these group programs provide a social worker or psychologist to help the family. Other groups provide vocational training for mothers and help with problems encountered at home such as proper nutrition and home economics. Group programs can be more formally organized into educational workshops for parents of handicapped children. These programs are more likely to include the child's father. Many center-based and home-based programs often include only the mother or the caretaker because they are offered during the day.

Home-based programs are most frequently available in rural areas although they are also found in urban areas. In this type of program the therapist travels from home to home to work with the parent and child on an individual basis. Advantages of the homebound program include being able to observe the child in a familiar environment, having the child receive individual attention, and regular participation as the parent and child do not have to travel to receive therapy. One disadvantage of the homebound program is that the family may still feel isolated, as no opportunity is provided for the parent or the child to meet other children with physical disabilities and their families.

The center-based program in which children are treated in groups with their parents present for part or all of the session, combined with an evening parent-education program and regularly scheduled home visits seems to be the ideal type of program.

ROLE DELINEATION

Parents frequently ask what the difference is between occupational and physical therapy. Although physical therapy and occupational therapy are two different disciplines, they frequently complement, usually overlap, and occasionally assume the other service's role depending on the facility, its philosophy, the skills of the therapists involved, and whether both services are available. For example, in rural hospitals, physical therapy may be the only available service, while occupational therapy may be the only service in some programs for the mentally retarded. The staff of a pediatric clinical facility usually appreciate that physical or occupational therapists require different types of knowledge and training and encourage some flexibility in therapy roles to provide the highest quality treatment for the child.

If physical therapy and occupational therapy are both available in a center, each will have specific responsibilities for treatment that are unique to their own service while sharing other roles in certain areas. For example, one responsibility of physical therapy might be increasing muscle strength, but if weakness is affecting a child's eye-hand skills, a function usually designated to occupational therapy, then the occupational therapist may also incorporate strengthening into her program.

Table 8.1 illustrates the general role of the pediatric occupational and physical therapist working with neurologically impaired children. The roles listed in the middle column represent overlapping areas. Overlap is more apt to occur in the treatment of the infant or nonambulatory child with neurological problems. Both disciplines are involved in patient and family education. Each service should explain to parents how its therapy program relates to other services involved with their child. Before beginning treatment, parents should ask the physical therapist and occupational therapist to explain their roles and responsibilities. Parents frequently visit a therapy unit and notice that their child's therapist may have different duties with another child whose problems seem similar to that of their own child. The individual needs of each child, family needs, the therapist's areas of expertise, and availability of other therapy services will influence a therapist's role in each child's treatment program.

Table 8.1. Role model for the pediatric occupational and physical therapist working with children with developmental disabilities

Occupational therapy	Overlapping areas	Physical therapy
Splinting	Feeding	Ambulation
Psychological and social behavior	Range of motion	Posture—static and dynamic
Eye-hand coordination	Fine motor development	Bracing—trunk, legs
Play	Transfers	Exercise, including strengthening trunk, neck, limbs
Prevocational evaluation/training	Gross motor development	
Homemaking	Dressing/bathing	
Learning difficulties	Adaptive equipment	
	Strengthening	
	Patient/family education	
	Behavior management	

The physical and occupational therapists' roles will be discussed in general for children with cerebral palsy, myelomeningocele, neuromuscular disorders, and mental retardation, with particular attention to the management of the infant, the toddler, and the school-aged child. This chapter is intended to tell parents and professionals what they should expect of a therapist, to suggest some appropriate questions they should ask, and to indicate the kinds of services that can be offered to children with neurological handicaps.

THE THERAPIST'S EVALUATION

Any therapeutic program designed for a child should begin with an initial comprehensive evaluation. These evaluations should be repeated about once every three months to help the therapist, parent, and physician document the child's progress or identify problem areas. Repeated evaluations also provide information on how well the therapeutic intervention program is meeting the child's needs and how he is responding to therapy. The therapist can then modify the treatment program according to the child's progress or lack of progress as documented by the evaluation. It may be impossible to complete a comprehensive evaluation of an infant or preschool child in one or two sessions if the child is uncooperative or irritable due to conflicts in feeding and nap schedules or lack of familiarity with the environment and the therapist. As the infant or young child becomes more comfortable and scheduling difficulties are worked out and as the therapist begins to adjust to the child's behaviors, more information can be obtained. A screening evaluation may be done initially, and a more comprehensive evaluation follow.

In some facilities, therapists may be asked to make a judgment about a child's intellectual capabilities or educational needs; however, unless they have had the proper training, this is probably not desirable even if trained professionals are not immediately available. It would be better to refer the child to a more comprehensive facility for this part of his initial evaluation.

The therapist needs to know about the motor and mental development of the child. Can an infant demonstrate good head control when on his stomach or his back, or when sitting; can he bring his hands together to hold a toy; will he look at the toy; can he follow verbal directions; and how well does he communicate? When a child is unable to accomplish a task that is appropriate for his age, the therapist also needs to break down the prerequisite motor components for the task to understand what skills the child has, why he is unable to accomplish the task, and what skills he needs to be able to perform the task or advance to the next developmental level.

The therapist should routinely assess the child's posture in various positions (depending on the child's age and capabilities)—stomach, back, sitting, quadruped (on hands and knees) and standing. She will evaluate the child's ability to move his head, arms, legs and trunk, his ability to roll, assume and change positions, walk, and run. The therapist is not only interested in whether the child accomplished the motor task, but how he did so. Were the movement patterns used normal or abnormal (if abnormal, could they interfere with further development)? How much time did it take to perform the task? What amount of effort did the child have to exert? Many of these findings are subjective; therefore, the therapist must be as precise and complete as possible in describing the movement patterns and posture of the child. Many facilities now document the children's posture and movement through the use of photographs, videotapes, or Super 8 movies. Gross and fine motor development, dressing and undressing, and self-care skills can be evaluated with the use of videotapes or by observation using standard developmental tests. Developmental reflexes and reactions should be assessed as a part of the evaluation of motor development. These tests may vary from one facility to another.

The musculature of the mouth; oral reflexes; structure of the mouth; coordination for sucking, swallowing, breathing, and chewing; lip, jaw, and tongue movements; and sensitivity of the mouth to temperature, touch, and textures should also be evaluated. The development of feeding skills and speech acquisition are associated with oral motor functions. Language and speech development are usually evaluated by a speech therapist; however, physical and occupational therapists frequently evaluate oral movements used for sucking, swallowing, and chewing. Physical and occupational therapists must be aware of a child's language capabilities in order to communicate with him and build good rapport.

The range of motion for the joints in the arms and legs is documented in degrees as active and passive movement. Active movement refers to the range of motion through which a joint can be moved by the child. Passive movement refers to the degrees through which the therapist can move the joint. It is necessary to check joint motion regularly so as to be aware of contractures or joint deformities the child may have and to document the effectiveness of treatment in decreasing or preventing limited or excessive joint motion. The child with restricted joint motion may not be able to stand or walk properly or reach for or transfer objects easily.

Sensory modalities that should be evaluated in young children include: response to light and deep touch; response to painful stimuli; proprioception (position of a body part in space); kinesthesia (sense of movement); response to visual stimuli, sounds, and vestibular stimuli. Vestibular stimuli test the balancing mechanism of the inner ear. Children with cerebral palsy may have impaired vision, hearing or deficits in other sensory areas. Sensory feedback helps a child learn how to repeat a movement.

Perhaps the most difficult evaluation the therapist performs, but the most essential as a basis of therapeutic management, is the assessment of movement patterns and influence of muscle tone. Muscle tone is measured according to distribution, type, and degree of influence on movement patterns. All children with cerebral palsy have abnormal patterns of movement that are affected by muscle tone. Muscle tone is classically examined by passively moving the arms and legs at each joint and noting the amount of resistance the muscles display to the passive stretch. The therapist evaluates changes in tone in response to stimulation such as noise, changes in position, and the amount of effort the child exerts to perform a task. Variations in muscle tone are observed in response to emotional changes, and may increase with excitement, irritability, or crying. Abnormal muscle tone will make it difficult for a child to move. Some aspects of the examination of muscle tone are assessed during the evaluation of posture and mobility, when testing reflexes, when evaluating gross and fine motor control, and when determining sensory acuity, as well as during tests of range of motion.

The most important aspect of evaluation is that the therapist provides parents with information about the child's condition and his progress. Parents should not hesitate to ask the therapist questions such as, "Is my child delayed in development? What areas of development are delayed (e.g., language, fine motor or gross motor development, cognitive development)? How will these problems affect my child in the future?" The therapist may not always have answers and may wish to correlate her findings with those of other professionals, but if parents ask questions they should be given all the available information.

At the conclusion of the evaluation, the therapist should be able to discuss the following:

1. The child's level of gross and fine motor development;
2. Normal and abnormal patterns of movement that the child displays;
3. Problems related to the child's muscle strength;
4. Limitations in the range of motion of joints as they relate to present disabilities and future problems that need to be avoided;
5. Sensory deficits and their effect on the child's development;
6. The child's behavior as it relates to his participation in a therapy group and his development.

Along with this information, the therapist should discuss the child's strengths and weaknesses as they relate to his present and future development. Although the therapist cannot predict a child's future with certainty because of the many physical, environmental, and psychological features which will influence development, she should be able to give parents some insight into the expectations she has for their child and to discuss the short- and long-term goals she thinks can be accomplished. Parents should be involved in planning these goals for their child so that they will understand why certain things need to be done and also so that they can judge their child's progress intelligently.

The information obtained from the evaluations gives the therapist a starting point or a basis for planning the child's treatment program. Each child should have an individual treatment program. There are several manuals and books available to the public on treatment of cerebral palsy, for example, but we urged parents to have the physical and occupational therapist select appropriate treatment activities and equipment individually for each child, to allow him to benefit from therapy maximally. One activity may be beneficial to one child but ineffective or potentially harmful to another.

PLANNING TREATMENT

Using information gathered from the evaluation procedures, physical and occupational therapists begin treatment planning by setting goals to reach. These should be stated precisely according to the motor ability to be achieved and the criteria for performance (such as time constraints or level of independence); they should be directed at functional activities; and parents should be included in planning from the start. An example of a treatment goal is: Robert will be able to move from lying on his back to the sitting position by himself within ten seconds.

When planning the treatment program, the therapist must be aware of the child's abilities and disabilities so that the planned activities are neither too easy to be challenging, nor too difficult and thus frustrating for the child. The social interaction between the infant and the caregiver (usually the mother) must also be considered in the planning phases of treatment. It is important that the father be included in the planning and in the home care of the child right from the start so that he becomes an active member of the team. Too often the contribution fathers can make is overlooked or not emphasized. This may lead to feelings of isolation and rejection on his part as the mother becomes more and more involved in the child's day-to-day care and has less time to spend with him. Fathers who share some of the child's daily care often design and build special equipment that makes it easier to take care of his child's physical needs and in some instances permits the child to be more self-sufficient.

Therapists have much to offer the handicapped child and his family. We want to emphasize again, however, that it is not as much as many parents desire or as much as some therapists claim is possible. We do not accept the idea that physical or occupational therapy is able to increase intelligence or accelerate the learning of school skills independent of motor control. Many children who have reading disabilities or are hyperactive may also be clumsy and have poor balance or eye-hand coordination. Formal programs that use the balance beam or emphasize visual control of motor function have not been shown to improve reading skills or lengthen the time a child can concentrate on doing his homework. The most that can be said is that by improving agility, the child may become more interested in sports, use some of his energies productively, and as a result receive the recognition every child needs in order to develop a feeling of self-worth. A family should expect a great deal from a well-run therapy program if their child has significant neuromuscular disabilities. They should not expect that any program of treatment designed by any medical professional will alter the basic structure of their child's brain.

REFERENCES

Barsch, R. H. 1965. The parent teacher partnership. The Council for Exceptional Children, Arlington, Virginia, p. 433.

Bobath, B. 1967. The very early treatment of cerebral palsy. Dev. Med. and Child Neurol., 9:373–390.

Bobath, B., and K. Bobath. 1975. Motor development in the different types of cerebral palsy. Heinemann Medical Books, London, p. 105.

Bobath, K. 1972. The motor deficit in patients with cerebral palsy. Spastics International Medical Publications, Lavenham Press, Lavenham, England, p. 54.

Brazelton, T. B. 1969. Infants and mothers. Dell Publishing, New York, p. 296.

Brazelton, T. B. 1973. Neonatal behavioral assessment scale. Spastics International Medical Publications, Lavenham Press, Lavenham, England, p. 61.

Caplan, F., ed. 1973. The first twelve months of life; your baby's growth month by month, 3rd ed. Grosset and Dunlap, New York, p. 255.

Connor, F. P., G. G. Williamson, and J. M. Siepp. 1978. Program guide for infants and toddlers with neuromotor and other developmental disabilities. Teachers College Press, New York, p. 415.

Cliff, F., J. Gray, and C. Nymann. 1977. Mothers can help. (A therapist's guide for formulating a developmental text for parents of special children), 2nd ed. Guyners Printing Co., El Paso, Texas, p. 212.

Finnie, N. R. 1975. Handling the young cerebral palsied child at home, 2nd ed. E. P. Dutton, New York, p. 224.

Frankenburg, W., J. Dobbs, and A. W. Fandal. 1970. Denver Developmental Screening Test Manual. University of Colorado Medical Center, Denver, p. 65.

Hainsworth, P., and M. Hainsworth. 1974. Pre-school screening system: Start of a longitudinal preventive approach. P.S.S., Pawtucket, R.I., p. 81.

Hainsworth, P., and M. Sigueland. 1969. Early identification of children with learning disabilities: The Meeting Street School screening test. Crippled Children and Adults of Rhode Island, Meeting Street School, Providence, R.I., p. 126.

Jones, S. 1976. Good things for babies. Houghton Mifflin, Boston, p. 115.

Milani-Comparetti, A. 1977. The Milani-Comparetti motor development screening test. Meyer Children's Rehabilitation Institute, Omaha, p. 30.

Pearson, P., and C. Williams, eds. 1972. Physical therapy services in the developmental disabilities. Charles C. Thomas, Springfield, Illinois, p. 460.

President's Committee on Mental Retardation. 1976. Changing patterns in residential services for the mentally retarded. U.S. Government Printing Office, Washington, D.C., p. 232.

Schafer, D. S., and M. S. Moersch, eds. 1977. Developmental programming for infants and young children, Vol. III. The University of Michigan Press, Ann Arbor, p. 144.

Twitchell, T. E. 1965. Attitudinal reflexes. J. of the Amer. Physical Therapy Assn., 45:411–418.

White, O., J. Minor, and B. Connolly, eds. 1977. A comprehensive handbook for management of children with developmental disabilities. University of Tennessee Center for the Health Sciences, Child Development Center, Memphis, p. 231.

Wolanski, N. 1975. Evaluation of motor development of infants. Polish Psychological Bulletin, 4:1.

9

THE SPEECH PATHOLOGIST

Many neurologically impaired children experience difficulty with communication. The problem may be mild and involve only the incorrect articulation of sounds, or it may be so severe that the child makes no attempts to communicate. Parents frequently report that difficulty in communicating is one of the most frustrating aspects of the day-to-day care of their handicapped child.

Before going on to discuss these problems, it is important that we clarify some terms for the purposes of this chapter. The broadest and perhaps most important term is *communication*. This will be defined as any interaction between two or more people that conveys a message. The interaction may involve the use of gestures, words, sounds, objects, or anything else that relays information. *Language* is a term used to refer to words and parts of words and the way words can be put together to have meaning. A child's language development includes his grasp of grammar, word meanings, and the length, complexity, and the completeness of his statements or questions. Language has two parts: (1) *comprehension*—what someone understands; and (2) *expression*—the person's use of words. It is important that comprehension and expression be considered separately when a child's use of language is judged. It should be remembered that language, as it is used in this chapter (words, talking, etc.) is only one way to communicate. A child's language refers to his ability to comprehend and to use words to communicate. If a child says words but does not use them to tell someone something, he is not communicating and is not using language. In this chapter *speech* will refer to the production of sounds—articulation. A child's speech is judged in terms of how clear or intelligible his words are and what sounds he uses or does not use. Any child whose speech is such that the listener is distracted from what the child is trying to say (his expressive language) by how he is saying it (his articulation or manner of production of words) is considered to have defective speech. The degree and importance of the defect may vary according to the amount of distraction it causes and the age of the child. A speech defect is considered significantly handicapping when the distraction it causes makes it difficult for him to communicate easily with others.

Parents should be concerned enough to seek help if:

1. Their child is one year to eighteen months old and does not seem to understand anything that is said to him without accompanying gestures (e.g., mother says "Daddy's coming" but the child does not respond. She points to the door and the child looks at the door to see who is coming).

2. A child is eighteen months old and does not attempt to communicate. He will not even use gestures (e.g., the child cries when he wants a cookie but does not point to it or pull someone to get it).

3. He is over two years old and is not using any words or only a few single words that do not consistently refer to something specific.

4. He is three years old and is easily confused when given directions (e.g., the child does not go get his shoes and socks when asked, but when his mother points to his bare feet he looks for them. Or he will get only one of the two or three objects requested).

5. He is 2½ to three years old and does not put words together to form short sentences.

6. He is three to 3½ years old and only family members can understand him.

7. He is 4½ to five years old with noticeably faulty sentence structure in which he omits pronouns, conjunctions, or prepositions (e.g., "Me go school," "I go school tomorrow," "Fall down hurt knee").

8. Parents are concerned about the child's communication skills, no matter how old he is.

9. The child seems embarrassed or self-conscious about his speech.

Parents who have decided they need professional advice about their child's language should contact their family doctor first. Although a doctor's referral is not usually required to obtain speech services, it is desirable because he can evaluate the child's progress in other areas of development. The doctor should be sensitive to parents' concerns and be able to refer them to an appropriate speech and hearing agency in the local community, or if the family lives in a rural area, to a facility in a nearby city. Many hospitals have speech pathologists on their staff, and there are community and private speech and hearing clinics available in most cities. Public school systems generally have facilities for the evaluation of speech and language development for preschool and school-aged children.

A hearing evaluation is extremely important if a child experiences difficulty learning or using language. It will probably be the first test performed after a physician's evaluation. Depending on the child's age and his ability to cooperate, his hearing can be tested by having him respond to or turn to noises of graded intensity. If the child cannot cooperate, it is still possible to evaluate his hearing using evoked-response audiometry. This involves putting the child to sleep and measuring his electrical response to noise. Hearing tests are sometimes done by a speech pathologist, but more often by physicians, physiologists, or trained technicians. The child will be referred to a speech pathologist who is associated with a hospital, clinic, or school after his level of hearing has been determined.

Most states require a speech pathologist to be certified or licensed in order to practice. Certification by the American Speech and Hearing Association and state licensing requires a master's degree and one year of supervised work experience. Those who are certified usually use the title "speech pathologist," "speech clinician," or "speech therapist." The title "speech correctionist" implies a bachelor's degree or less education and no certification or license. Many school systems do not require certification or licensing of personnel treating children with speech and language problems. It is important to inquire into the training and qualifications of a child's speech therapist. If the therapist is not certified or licensed and the child is making little progress, parents should consider an alternative program that has staff with more professional training.

The way in which a speech pathologist evaluates a child depends on his age and his language abilities. An evaluation will include pictures and games to elicit responses that require the use of specific language structures ("s" to indicate plural; "ed" to show past tense). The speech pathologist's goal is to get an estimate of the types of language structures the child understands and is able to use. The therapist need not hear every word or sentence the child can use to make such an estimate. The therapist will attempt to assess the child's communication level, including his response to and use of words as well as his responses to and use of nonverbal communication. A speech and language evaluation will proceed more smoothly if the parent does not attempt to "prepare" the child beforehand by urging him to "Talk for the lady," etc. If there is a need to "explain" this visit to a child, it is best to tell him he is going to get to see some pictures he will like. The parent's approach must be matter-of-fact so that the child will be relaxed.

Although parents may not be in the room during the evaluation, they should expect to have the results explained to them after the testing is complete. The results are usually best explained by the speech pathologist. Some doctors prefer that a report be sent to them so that they can then discuss the findings in relation to the child's overall development. However, parents may ask to speak to the speech pathologist after talking with the doctor. The speech therapist will usually not discuss causes for a child's handicap but will be able to give specific information about the child's speech and language development and areas of weakness.

Recommendations regarding the need for therapy will depend on the results of the evaluation. A child may not need speech or language therapy. This will be true if he is using speech and language at the expected levels for his age, or his speech and language development is progressing at a rate consistent with his other developmental parameters. If the child does need help with his speech or language, the extent of this help will be based on the severity of his problem relative to his overall intellectual capabilities. The milder the child's problem, the less intense the therapy will be. Most public schools, hospitals, and clinics offer speech and/or language therapy. One-half-hour to one-hour sessions, one or more times per week, are generally prescribed. Therapy is usually not related to the child's school or preschool placement and may be done in his free time rather than during school hours. This type of therapy is probably best done on an individual basis or at most in a group of two or three. The therapist should encourage parents to observe therapy sessions and should provide parents with material and instructions so that they can continue to work with the child at home. However, parents should not be put in the position of "teaching" new sounds or language concepts. Instead, they should provide practice for a sound or concept that has been taught to the child in his therapy sessions.

If the child's communication problem is likely to keep him from profiting from school placement, enrollment in a half- or full-day class for language- or speech-impaired children may be recommended. This class is usually taught by a speech pathologist or has a speech pathologist involved in it on a full-time basis. The class generally tries to accomplish what any classroom accomplishes, but in the process works intensely on a child's speech and language skills. There should be no more than ten children in such a class, and parents should be encouraged to become frequent observers.

If a child's primary problem is significant mental retardation or physical disability rather than a pure speech or language defect, appropriate placement for the primary problem is necessary. A speech pathologist should be available to offer suggestions to the teacher and to work with the child. If the child's primary problem is so severe as to negate or minimize the chances of intelligible, useful oral language, alternative communication methods such as sign language may be necessary and can be taught by the speech pathologist.

It is difficult, and in most instances impossible, to know what the end result of therapy will be. That, however, does not mean that a therapist will not have short- and long-term goals. The parents should know what the goals are. It is likely that the goals will have to be modified somewhat as the therapist gets to know the child's abilities better. These changes should be explained to the family so that they do not worry. Aside from being aware of the goals for their child, parents have a right to know what improvement is taking place. If a therapist is not willing to share this information and/or the family notices no improvement, they are justified in seeking services elsewhere.

It seems inevitable that relatives, friends, and often strangers on the street become self-appointed experts when a child does not talk as expected. They are generally more than willing to tell parents what they have done wrong, what the child does wrong, and, of course, what should be done about it. Along with the problems associated with such encounters, we have found that when a child does not talk as much as a parent or some other adult believes he should, several events can take place that may further delay the child's language development. The natural tendency when a child doesn't talk much is for those around him to talk less to him. Adults must work to counteract this tendency, since children delayed in learning language need *more* opportunities than others to hear language, *not less.* Another common reaction is for parents to increase their efforts to make the child talk by making him repeat words he already can say. Parents often find themselves saying "What's this?," "Say———," "Tell me———." The result is that the child is constantly being tested and he has little chance to learn new words and structures. A child will never use a word or group of words consistently unless he understands what they mean. Thus efforts to improve a child's language should focus on the parent's use of meaningful words, from which the child should learn to understand and use the words himself.

The following are some suggestions that may be given to parents so that they can provide an overall stimulating language environment for a child, with a few suggestions for specific activities that parents and child might enjoy. They are useful suggestions for parents of children whose language is developing normally as well as those parents who have a child with delayed language development or poor speech.

However, they are an adjunct to therapy and not a substitute for it. A child's speech therapist may wish to modify these suggestions depending on the exact nature of the child's speech or language problem.

1. *Talk to your child*
2. *When to talk*

   Talk to your child when you can get his attention. If your child is not paying attention to you, then it is probable that your talking to him is not helping him to learn new language. It is not necessary or desirable to keep up a running monologue in which the child has no interest.

3. *What to talk about*

   a. Talk about what your child is doing. When you and your child are together, you can describe what your child is doing. As he plays, you can describe what he is building with the blocks, what he is riding on, toys he is playing with, etc. You can make comments about your child's actions, e.g., "Irving, you drank all your milk. It is all gone!" "Oh look, the truck is going under the chair." etc.

   b. Talk about what you are doing. As you do things with your child, you can describe your own actions and, when appropriate, your feelings, e.g., "I'll make you a sandwich. Here's the bread. Here's the peanut butter."

   c. Talk about things in the child's environment. You can describe the child's toys, clothing, what you see when you take a walk or go for a drive in the car or when you go to the market or to the zoo, e.g., "Look at the pretty flower." "Let's put some apples in the cart."

4. *How to talk*

   a. Talk at a rate at which your child can understand. This will probably be a little slower than the rate you use when talking to another adult.

   b. Enunciate your words clearly so that you will be giving your child a good, clear speech model. Do not over-exaggerate your mouth movements, however.

   c. Keep the language simple enough for the child to be able to understand what you are talking about. Talk at a level at which your child can understand some of what is said, but above the level at which he can understand all of what is said.

   d. Look for signs of understanding or confusion. When your child understands a particular word, you should begin using that word in simple sentences. When your child understands simple sentences, you should begin rephrasing the ideas in more complex sentences, longer sentences, or in other words. Learn to look for signs of comprehension from your child. If your child seems confused, use whatever means you have to make yourself understood. The more your child understands, the more willing and eager he will be to pay attention and try to understand still more. Confusion may force him to give up trying to understand at all. However, gestures should be used to amplify or reinforce verbal communication and not as a substitute for these skills, e.g., if the child knows "ball," you should begin using, "Bounce the ball." "The ball is big," etc.; then move to, "Throw the ball to me." "Bounce the ball up and down." "That's a big ball."

   e. Do not use "baby talk" or mispronounce words on purpose. Even if you are speaking in short phrases or sentences, it will only retard your child's language development.

5. *Listen to your child*

   It can be very reassuring to a young child to have an adult listen intently to what he is saying. Take time to pay attention to your child's utterances and other attempts to communicate and try to figure out what he is trying to "say." Just the fact that he is getting your undivided attention may make it worthwhile for your child to attempt to communicate with you. The more effort you make to understand what your child is trying to tell you, the harder he will try to make what he is "saying" comprehensible.

6. *Converse with your child*

   Conversation implies an exchange of thoughts or ideas. It implies that you say something and then your child responds, or the other way around. Therefore, conversing with your child means that you respond to what he says and you expect him to respond to what you say. There is a "give and take" implied in conversation. You are alternately both the talker and the listener. Conversing with your child is an important way for you to help him learn language and to learn to converse. Do not let talking become a one-way street.

7. *Motivate your child to communicate and then help him to communicate verbally*

   a. Your child must have something to "say."
   b. He must try to "say" it (although this may not be with words at the early stages).
   c. Then you can provide him with the words for saying it.

Usually you only have to wait long enough for the child to have something to say. You may know your child so well that you know things he wants and needs even before he does. Often, in the interest of saving time and keeping things running smoothly, you take care of all his needs and desires for him without giving him a chance to recognize that he has a need, and then figure out a way to communicate that need to you. If, however, *you refrain from anticipating your child's needs* and let him tell you himself, in some way, then you can supply him with appropriate words for what he wants. For example, sometimes when serving ice cream to your child you might "forget" to give him a spoon. This creates an opportunity for your child to have to communicate that to you. When he lets you know something is wrong, you can give him a model for what he wants to say ("Oh, a spoon! Mommy forgot your spoon. Here's a spoon. Now you can eat your ice cream!")

Provide good speech and language models for your child, but do not criticize him for falling short of the model. *No correction should be made of his attempts to talk at any time.* Encourage him to talk, but do not pressure him to do so.

8. *Use pictures as an aid in talking to your child*

You may wish to make a scrapbook of pictures or make individual picture cards of familiar items in your child's world, e.g., foods he eats, clothing he wears, places he often goes, toys he likes to play with, friends and family members, things he likes to do, etc.

Such pictures provide a tool for talking with your child about things of interest to him and will help you talk about such things in a very meaningful way.

When looking at pictures or when talking about anything with your child, do not try to force your child to talk. If your child does try to tell you something, do everything you can to figure out what he is trying to say. Encourage him to use pictures, pointing, gestures, or anything else to get his idea across. Let him know that you are interested enough in what he has to say to make an effort yourself to understand him. Avoid saying things like "You don't talk right," "I never understand anything you say," etc. Instead, say things like "Show me," "Do you mean ———?," "Is this what you want?," etc.

9. *Read to your child*

Young children like books with big pictures and small amounts of print. Sometimes children prefer simply to look at the pictures and talk about them. If this is how your child wants to use books, then this is how you should use books with him. Some children enjoy hearing the same story over and over and over. If your child likes to hear the same story, then you should comply with his wishes, but try to introduce new stories along with his old favorites.

Pick a quiet area and a quiet time of day for reading to your child. Spend five to ten minutes every day looking at books or reading to your child. Take your child with you to the library and let him help choose the books.

Talking is a social process and it should be pleasant. If it is not a pleasant experience, why would any child want to do it? If parents are frustrated in their attempts to talk with their child or the child seems frustrated, parents should seek further guidance quickly.

REFERENCES

Battin, R. R., and C. D. Haug. 1978. Speech and language delay: A home training program, 3rd ed. Charles C. Thomas, Springfield, Ill., p. 77.

Craft, M. 1969. Speech delay: its treatment by speech play. Williams and Wilkins, Baltimore, p. 102.

Espir, M. L. E., and F. C. Rose. 1970. The basic neurology of speech. F. A. Davis, Philadelphia, p. 134.

Hopper, R., and R. Naremore. 1973. Children's speech: A practical introduction to communication development. Harper and Row, New York, p. 140.

Irwin, O. C. 1972. Communication variables of cerebral palsied and mentally retarded children. Charles C. Thomas, Springfield, Ill., p. 370.

Kleffner, F. R. 1973. Language disorders in children. Bobbs-Merrill, Boston, p. 60.

Mysak, E. D. 1976. Pathologies of speech systems. Williams and Wilkins, Baltimore, p. 297.

Oratio, A. R. 1977. Supervision in speech pathology: Handbook for supervisors and clinicians. University Park Press, Baltimore, p. 148.

Renfrew, C. E. 1972. Speech disorders in children. Pergamon Press, New York, p. 69.

10

THE SOCIAL WORKER

Raising a handicapped child can be a difficult experience because of the repeated frustrations the child and his family encounter. These often occur when the handicap is first recognized. Parents must then suddenly confront and cope with intense and conflicting feelings about themselves and their child. At the same time, these families must learn to find their way through a complex and confusing maze of bureaucracies that exist in hospitals, clinics, and in the private and public agencies that administer programs for the handicapped.

The responsibilities of the social worker are twofold: (1) to help parents deal with their feelings so that they can plan intelligently and objectively for their child's future; and (2) to teach parents how to use the resources in their community effectively to secure assistance for their child and their family. Efficient use of social services can help parents free the time and energy they will need to devote to their child, their family, and themselves.

Social workers offer a wide range of services. These include support in coping with stress, help in dealing with medical personnel, and assistance in locating appropriate resources for practical concerns, such as financial arrangements, child care, transportation, and housing. Social workers are employed in hospitals, crippled children's programs, schools, residential treatment facilities, adoption agencies, child guidance and psychiatric clinics, agencies such as the Association for Retarded Citizens and the Epilepsy Foundation, as well as in private practice.

SOCIAL WORK SPECIALTIES

Most social workers are trained to one of two levels: the Master of Social Work (M.S.W.) or the Bachelor of Social Work (B.S.W.). A social worker with a master's degree has had specific training in family, group, and individual therapy and counseling. They are also familiar with health care systems, social welfare agencies and educational programs. Although curriculums at different schools vary, all M.S.W. social workers have completed a practicum in which they work under supervision in an agency for a prescribed number of hours each week. The purpose of the practicum is to give each student the opportunity to work directly with families under the careful direction and supervision of a social worker who is a senior member of the agency staff.

Social workers with a bachelor's degree (B.S.W.) usually have not had the advanced training to function as therapists; however, their training does include most other aspects of social work. Social workers with a bachelor's degree are supervised by those with master's degrees. They receive extensive on-the-job training and should have strong back-up available from senior social workers.

Caseworkers are individuals employed by agencies to gather information about clients and their families to determine initial and continuing eligibility for assistance. They may have no professional training but at the discretion of the agency, they may perform all the functions of a trained social worker, including therapy and counseling.

Social workers are often categorized by the area in which they work as well as by their educational background. Medical social workers are employed by hospitals or outpatient clinics. Psychiatric social workers are usually part of the staff of a child guidance clinic, psychiatric clinic, or psychiatric hospital. Residential treatment facilities often have psychiatric and medical social workers on their staff. Educational social workers are typically employed by schools and by government agencies concerned with vocational and educational training. Various government agencies, such as the crippled children's programs, will often employ social workers to relay family needs to them. As noted earlier, social workers are part of the staff of numerous private agencies concerned with the needs of handicapped children. Social workers engaged in private practice usually limit their interest to special areas such as family or individual therapy. Some medical insurance plans do not cover fees charged by social workers in private practice. Most social workers attached to hospitals, schools, or other semipublic or public agencies do not charge families a fee for their services. Generally, these people are paid a fixed salary by the agency for which they work.

A common misconception is that only the poor are referred for social services. This notion probably derives from the historical development of social work. Initially social workers were volunteers whose primary interest and function was to find money and goods for needy families. Although this is certainly an essential part of helping a family with a handicapped child, it actually represents only one aspect of the responsibilities assumed by modern social workers. In most settings social workers function as members of a team. They act as the parents' ally and as the child's advocate. They help identify the needs of the child and his family and discuss these needs and the family's concerns with other members of the team.

Since social workers provide a wide variety of services, families from all socioeconomic levels are referred to them, and these referrals come from many sources. Depending on the child's age at the time of diagnosis and the nature of his neurological disability, a family may be asked to see several social workers over the course of the child's development. Often these visits are for different purposes. For example, if the child has a significant birth defect or serious neurological damage that is evident at birth, the physician or some other member of the hospital medical staff should ask that a social worker talk with the parents as soon as possible in order to help them through the initial period of acute distress. Later, after the initial grief subsides, the social worker will begin to help the family organize plans for the child's care when he is discharged from the hospital. Similarly, if the child's neurological problem suddenly worsens (for example, he may need to enter an intensive care unit because of continuous epileptic seizures), a social worker may be asked to counsel the family.

A teacher may refer a child and his family to a school social worker because she feels the child has behavioral or emotional problems that require counseling. She also may have noticed that the child's state of nutrition or dress suggests that the family needs the aid of one or more local agencies. A child's physician or teacher might refer a family to a Community Mental Health Center where it is possible that care will be given by a psychiatric social worker. Parents may seek aid for their child from one or more private or government agencies, where their problems will be evaluated by a staff social worker, who will gather information to help determine their eligibility for services.

CONFIDENTIALITY

While parents have the right to see a social worker, they also have the right *not* to see a social worker (with the exception of Protective Services, which pertains to child abuse and neglect). Parents also have the right, if they wish, to limit the kind of aid they are willing to accept from a social worker or the agency they represent. This is certainly the case when parents are asked to see a medical, school, or psychiatric social worker. However, when dealing with a public or private agency, parents will probably not be provided with services if they severely restrict the social worker's investigation of the family's circumstances. If parents feel that their application for services has been rejected for what appears to be inadequate reasons, they can obtain a copy of the social worker's written report under the Freedom of Information Act. This report should set forth in detail the reasons why the family was not considered eligible for aid. Parents may appeal a decision they consider unfair by following the procedures required by the agency involved. Each agency should have procedures for appealing a denied petition. Most, if not all, agencies have specific procedures that regulate appeal procedures before they will reconsider a decision. Parents can contact an agency's director of social services for information regarding the regulations that govern an appeal.

Another issue that may arise is the confidentiality of any information a family gives to a social worker. All information provided to a social worker by parents, family members, or the legal guardian of a child is considered confidential and must be treated accordingly. No social worker can release information about a child or a family to another hospital, clinic, or agency without written permission from the parents or the child's legal guardian. Parents have the right (and the obligation) to read any reports concerning them or any member of their family before they grant permission to transmit information to another agency.

However, within a particular unit, such as a hospital, clinic, school, or agency, social workers are part of a team, and it is their responsibility to relay information about a child and its family to other team members. Since the social worker is the child's and the family's advocate, this is generally done in the most favorable terms possible. If there is certain information that parents do not wish recorded, or even discussed with other agency members, they should make certain that this is clear to the social worker at the time of the conversation so that if there is any question about withholding information, the issue can be resolved at that time. Social workers have a dual responsibility, both to the client and to the institution that employs them. Realistically, there are limits to the amount and the nature of pertinent information they can withhold from other team members even if it is highly personal. For example, if a parent confided to the medical social worker that he was seriously contemplating suicide because of the problems associated with his child's handicap, the social worker would be obligated to persuade the client to seek help and to communicate information about the client's mood to the family physician. Similarly, if a client were applying for public or private financial aid, an agency social worker could not be expected to withhold information about income resources, even if this had been told to them in strict confidence.

SPECIFIC SOCIAL WORK SERVICES

In the following section we will discuss five specific services a social worker can offer: providing emotional support, locating sources of financial assistance, coordinating community services, aiding in placement planning, and initiating referrals to appropriate agencies.

Emotional support

A major focus of social work intervention is providing emotional support to families and children. The degree of involvement and depth of intervention depend on training, caseload, and other professional back-up. Some social workers have taken courses and worked under the supervision of a senior worker in child therapy, family counseling, marriage counseling, and other modes of treatment. Many of these workers take postgraduate courses as well as continued training through the institution for which they work.

A social worker with a heavy caseload may well have to refer a family to someone else in the agency or to another agency if he or she is too busy to meet the family's needs. For instance, the social worker with a caseload of three hundred might not have the time for long-term individual therapy; however, this type of care might be the main function of a social worker in a child guidance center that is geared to a small caseload.

As has been mentioned before, social workers are team members, and their function depends in part on the other professionals included in the team. For example, in a school, a social worker would be working with guidance personnel, teachers, and administrators and would not treat a child with a serious psychiatric disorder. However, a social worker employed in an outpatient psychiatry unit who works with psychiatrists and psychologists might well be asked to take over some aspect of the treatment of a child with a serious behavioral problem, since she has appropriate professionals to consult with.

In many instances, the support given by social workers, psychiatrists, or psychologists is not significantly different. The major differences between the three professions are as follows: The psychiatrist is a medical doctor who can prescribe medicine and other treatment (such as electroshock) and who has admitting privileges to a hospital. Psychiatrists have special training in both long-term and short-term therapies. They do not generally provide practical services such as help with child care, educational placement, and financial assistance. Psychologists have special training in psychotherapy, the administration and interpretation of psychological tests, and counseling. Many are also trained in hypnosis, relaxation techniques, and behavior modification. The social worker does not prescribe medication or administer tests; however, the social worker has a broad knowledge of the community and its resources. These distinctions may vary from region to region and from agency to agency. Families will find broad differences in their experiences.

The social worker is frequently the "front line" mental health worker. As such, she will take a preventive approach to mental health and social problems. She is skilled in the identification of major and minor disturbances and may be charged with the responsibility of providing treatment for these problems either directly or by referral.

Many families are stunned by their reactions to their handicapped children and need a sounding board to discuss these reactions. Most people will find that their reactions are, in fact, not uncommon. A social worker can help parents sort through these troublesome new emotions.

Social workers are well acquainted with symptoms of problems that can fester if not dealt with promptly. Signs of marital stress, problems with siblings, and prolonged guilt and grieving are some examples. Prompt intervention, whether it be provision of a temporary homemaker, financial counseling, or short-term therapy, can frequently make the difference between an early resolution or prolonged suffering and the need for more intense professional help later.

Finances

Raising a handicapped child can be financially devastating. One of the most productive functions of a social worker is to help families locate resources and programs that can provide financial assistance. There are numerous programs available to help neurologically impaired children. These programs vary greatly from state to state, as well as in communities within each state. Eligibility for assistance may vary according to the nature and severity of the disability and any restrictions placed upon public and private programs by individual communities.

A family can pay for a child's expenses in one or more of the following ways: (1) private insurance programs; (2) Medicaid; (3) Crippled Children's Programs; (4) funds from private agencies such as the Muscular Dystrophy Association or the National Foundation (March of Dimes), etc.; or (5) by personal funds. There is also a federal program known as Supplemental Security Income that provides a financial grant as well as services such as assistance with day care, a homemaker, and transportation for families of handicapped children.

The role of the social worker in helping with financial planning is to evaluate a family's resources and needs. The financial demands made on a family will depend on the child's age and the nature of his handicap. The social worker may make a family aware of important questions about their child's health now and in the future that will directly relate to the financial assistance that will be needed. The social worker can also suggest questions that should be discussed with the child's doctor or even contact the doctor to obtain the information for the parents. Questions that may be extremely important regarding the financial future of a family may not be of immediate concern to the doctor but they are important for parents to think about early in a child's life. For example, it may be important to know whether the child will need repeated diagnostic tests in the future (such as laboratory studies to measure levels of antiepileptic drugs) two, three, four or more times each year. Will it be necessary for the child to be hospitalized for more complicated radiological evaluations? Will he need expensive dental work? Is it possible that the child will require a specially outfitted wheelchair or other expensive appliances? Will the parents need to make changes in their apartment or home to provide for the use of a wheelchair? Will a hospital bed be needed? A physician may not be able to give precise answers to each of these questions, particularly if they involve planning for some years in the future. However, he can give a general picture that will have meaning to the social worker in helping to plan for a family's financial future.

Private insurance The most common type of health coverage is the private insurance program. This includes companies such as Blue Cross/Blue Shield, Prudential, Travelers, etc. These programs are generally available through a parent's place of employment. Many employers will pay the cost of an employee's insurance and will offer a family plan that can be purchased through a payroll deduction system. However, children born with neurological defects or who already have a neurological handicap sometimes will not be covered if the parents enter the plan after the handicap is known. It is particularly important for parents of a handicapped child to compare programs when contemplating a change in jobs. The extent to which a handicapped child is covered by medical insurance may be a key factor in determining whether a parent should accept a new position with another company. Many companies offer insurance programs that will cover neurological disabilities present at birth, and others offer optional programs at an extra premium that will continue to cover children who have a neurological disability. If there is a choice of several policies, and the parents have difficulty interpreting how these policies relate to their child, they would be wise to discuss the matter with an insurance broker or a social worker.

In recent years private organizations, such as the Epilepsy Foundation of America, the National Association for Retarded Children, and United Cerebral Palsy have lobbied extensively and successfully to try to obtain more reasonable insurance rates for children and adults who suffer from neurological handicaps. It is frequently possible for children with handicaps to obtain life insurance as well as medical insurance.

In evaluating a private medical insurance policy, parents should obtain the following information:

1. What is the lifetime limit of the policy?

2. Does it cover physician's fees as well as the expense of hospitalization?

3. Does it cover outpatient laboratory fees and, if so, will it cover fees for the same type laboratory tests if they have to be performed several times in a year?

4. Can the policy be continued if employment with a particular company or association with a particular group ceases?

5. Is there a deductible clause in which a family will have to pay the first $100 or $200 of their medical fees each year and a certain percentage of all fees thereafter? Is there a limit on the amount of money that the policy will cover during any one year? For example, if a child should have to be hospitalized for a long period of time for an extensive orthopedic surgical procedure, will these expenses be covered after $5,000–$10,000 of hospital fees have been expended by the insurance company?

These and many other questions should certainly be uppermost in a family's mind when considering an insurance policy to cover the family and their handicapped child. Again, the wisest approach, we feel, is to consult a knowledgeable social worker or insurance agent before giving up any policy that will cover a neurological handicap.

Medicaid Medicaid, also known as Title XIX Medical Assistance, is a combined state and federal program that has eligibility requirements and benefits that vary markedly from state to state. Several factors determine whether a family is eligible for the program, and since the variability is so great, they cannot be specified for each area of the country. In general, eligibility is determined by the family's income, the value of their possessions, the number of children in the family, the number of disabled children in the family, and the nature and extent of the neurological disability. Medicaid pays for care that is received from hospitals, clinics, and private physicians. However, it is well to be aware that a private hospital, a clinic, or a physician need not accept Medicaid as payment. Many physicians do not accept Medicaid patients because of the immense amount of paperwork involved in obtaining a fee and the relatively low fee for the amount of work expended compared with that which they can obtain from private practice. Therefore, if the only way a family can cover their expenses is by Medicaid, they may find that they have a limited choice of hospitals, clinics, and physicians. Nevertheless, many excellent specialists and hospitals will accept Medicaid in full payment of their bill. Medicaid also covers medication and equipment such as wheelchairs or crutches. The social worker, physician, and physical therapist must work together to decide which aids are needed and to assist in ordering these items, because in most states Medicaid will only cover medication prescribed by a physician and equipment if the order is signed by a doctor. When a family is considering whether to apply for Medicaid, a social worker can be of great help. She can help decide if a family is eligible, for instance, by determining whether outstanding medical bills are sufficiently large to overcome discrepancies in a family's income that might otherwise make them ineligible for the program. She can also indicate what services in the community will be available to a family and which community hospitals will accept Medicaid as payment.

The address of the nearest Medicaid office can be found in the telephone directory under the city or county government's listings. No referral is needed to apply for services.

Crippled Children's programs Crippled Children's Services are a system of federal and state funded programs that were established to provide quality medical care to chronically ill children. The programs are administered by each state and the coverage, again, varies greatly from state to state. For example, in some states both the diagnosis and treatment of seizure disorders are fully covered, including all outpatient medications. In other states, no medication and only one visit to the doctor for a diagnosis may be covered. Some states will cover only diseases in which the patient can be restored to a full, active, and productive life in the community. Other states interpret the program in a much broader sense and cover many chronic diseases in which only a partial recovery is possible. A medical or agency social worker will be able to tell a family if their child's illness is covered in that state.

There are also financial criteria for receiving Crippled Children's funds. This too varies from state to state; when in doubt, it always makes sense to apply. Once again, a family should be sure to include any outstanding medical bills with their application, since this will affect eligibility for the program despite family income.

If it is determined that a family is not eligible for aid from a Crippled Children's program at a local level, parents can appeal that decision. The procedure for this varies from state to state. A parent should call or write the Crippled Children's office and request information on how to appeal a decision. Furthermore, if they find that their child's disease is not covered by the Crippled Children's program, they can petition for a change in coverage. Specific diseases which a Crippled Children's program will cover are determined at the state level. Change is a difficult process that requires hard work on the part of many families who have children with similar illnesses. Often social service agencies, as well as social workers from private agencies with an interest in specific neurological disabilities, have helped families to work for change and have been successful. The most obvious example of this is the inclusion of mentally retarded children in some state Crippled Children's programs. Political action is required to make such a change. Families can begin by writing letters to their state senator or assemblyman, and by enlisting the aid of other families who have children with a similar problem. Groups of families in each community must be organized in order to effect change. This is often best done through private organizations in the community interested in the child's disease. It is often helpful to enlist the aid of the family physician in such efforts, since he can describe your child's problems in terms that can be appreciated by other members of the medical community. Unfortunately, in many states the bureaucracy of Crippled Children's programs is relatively inflexible, and any application for substantial changes in services covered must be done through the state legislature.

Supplemental Security Income (SSI)   This is a program that is federally funded through the Social Security Administration. This program was initiated to aid families with children as well as adults with long-term, chronic illnesses. A child may be eligible as soon as a diagnosis is made, even if this is in the first days of life. Many people are confused as to the qualifications for eligibility, assuming it is determined by the social security status of the wage earner in the family. This is not true. There are two criteria for eligibility: the financial resources of the family and the nature of the child's medical condition. When applying for this type of financial assistance, the family again should include outstanding medical bills. To establish medical eligibility, parents should have a letter from their physician that includes the diagnosis, prognosis, an exact summary of the child's present disabilities, the disabilities he anticipates in the future, and current medications. Parents seeking aid should make application in person at the nearest Social Security office. Parents are advised to contact the office in advance and request a complete list of the documents they will need to bring with them. These include rent receipts, the child's social security number, and his birth certificate.

If a family is refused assistance on medical grounds by the Supplemental Security Income Program, they should always file an appeal to have the decision reviewed. Printed information about how to appeal is available through the nearest Social Security office. Frequently, if there is additional pertinent information, the refusal is reversed on appeal.

Basically, Supplemental Security Income pays a lump sum per month to help with the added expense of raising a handicapped child. The amount of money will depend on the family's income. If a family is eligible for even one dollar per month, they are also eligible for Medicaid. Parents often assume that Medicaid will begin automatically because they are receiving SSI. This is not always so. Sometimes parents must make separate application for benefits at their nearest Medicaid office.

There are many private agencies interested in helping families who have children with specific neurological disabilities. The national headquarters of some of the major agencies are listed in the appendix to this chapter. Many of these agencies will have local chapters that can be found in the telephone directory. In other instances, parents must write to the national headquarters to find out if a local chapter exists and if their child is eligible for assistance.

Private agencies have a number of functions. They may pay some of the expenses involved in a child's medical care. Perhaps the best example of this is the Muscular Dystrophy Association, which spends a great deal of its income each year on providing diagnosis and treatment for children having or suspected of having neuromuscular disorders. In other instances, these agencies might not pay for direct care, but they might pay for a specific piece of equipment that parents are otherwise unable to obtain. These agencies will also direct families to clinics and physicians in the community who have special competence in the area of a child's disability. They will also put parents in contact with other families who have children with similar disabilities. These groups are often very helpful because they provide opportunities for parents to share useful information with others. They are also a source of emotional support and as a group they can work effectively to achieve changes in medical care systems. Frequently these agencies employ social service workers to help with these matters and to work with a family and their child on a day-to-day basis. Lastly, these agencies are often important sources of funds for research into neurological disabilities that affect children.

Coordination of community services

Often both parents of a handicapped child must work in order to earn the extra income necessary to pay for the child's medical care. If both parents are employed, the day-to-day care of an impaired child is a serious problem for the family. There are community resources that can sometimes be made available to help families with this problem. After evaluating a family's situation and a child's needs, a social worker can determine whether the family is eligible for the services of a part-time homemaker and through what program these services might be obtained. She should also be able to locate community daycare centers that will accept handicapped children, advise of their locations, and help families with enrollment procedures. She will know whether transportation to and from such centers is available or whether it must be provided by the family. She should know a good deal about the nursery schools in the community. For example, she should know if they take children with handicaps, how old the child must be to enroll, and whether they require that the child be toilet-trained before being accepted in the school. Many nursery schools and even some day-care centers will not enroll a child who is not toilet-trained. The social worker should know the fees for such programs and whether a family is eligible to have part or all of the fees paid by a public agency. Often a social worker will be able to help a family with less formal arrangements. For example, it may be possible for a child to be cared for by a neighbor while mother works, or there may be a relative in the family who is willing to assume that responsibility.

Parents are often reluctant to ask for such help, whereas a social worker can frequently make these arrangements easier for everyone involved.

As a child grows older, education will become an increasingly important issue. Education is the responsibility of the school system (see Chapter 7); however, either a medical social worker, a school social worker, or an agency social worker can still be of considerable help to a family. They can help obtain proper equipment to allow a child to be sufficiently mobile to attend classes, or can arrange transportation to and from school. A social worker can help to see that such services are made available to a family and to contact community agencies on behalf of the child. Frequently, the school day will be over before a mother has finished work and there will be no one at home to care for the child. There is then the question of who is to care for the child until the mother returns home. The social worker can direct a family to an appropriate day care facility in the community that will assume this responsibility. If such facilities are not satisfactory or do not exist, she may be able to help a family enlist the aid of neighbors.

While many children with neurological disabilities improve or remain stable as they grow older, others worsen and may be forced to remain at home. If this is the case, the educational system in most communities will provide a home teacher for the child. A social worker can help families apply for this type of service for their child and tell parents what type of medical information they will have to get from their doctor in order to obtain such services.

Placement planning

Short-term placement When a family has a child with a serious handicap that requires an enormous amount of attention, it is not unusual for parents to wish for brief periods of relief from their constant responsibility. There is certainly nothing wrong with this; in fact, it is generally better for both the family and the child that such "vacations" be taken on a relatively regular basis. This means that the child must be cared for by someone else for a short period of time, i.e., one or two weeks. Such placements can be arranged in close-knit families among family members or, in close-knit neighborhoods, by neighbors. If parents are active in an organization such as United Cerebral Palsy or the Muscular Dystrophy Association, they may find other members of their local chapter who are willing to help care for their child in exchange for similar services when they need a vacation. There are other community resources available that a social worker can locate for parents. Institutions will take handicapped children for limited periods of time if arrangements are made well in advance. In addition, many agencies have camps for handicapped children that operate during the summer. These camps are usually limited to children with a particular type of handicap, such as muscular dystrophy or cerebral palsy. They are run by well-trained counselors and usually have an excellent medical staff immediately available. These camps not only provide an opportunity for parents to have a vacation, but they are a superb opportunity for the child to participate in supervised games and physical activities that might not otherwise be available to him. A knowledgeable social worker will know if such resources exist.

Occasionally, it is necessary to place a handicapped child in an institutional setting for a longer period of time because of medical or emotional problems. Usually these are children with severe seizure disorders or serious behavioral problems who do not require hospitalization but who do require a highly structured environment. This type of environment may be geared to meet their medical needs or to offer a consistent approach by which their behavior can be modified. This is not a permanent placement; many of these children are able to return home once they have improved sufficiently. Such programs are often run by the state mental health department or by private agencies. The institutions that offer such programs may be located some distance away, but they may still be available to any child needing their services. A social worker should know where such programs are located and whether a state agency or a private agency will pay all or part of the fee.

Long-term placement In some instances it is necessary to place a handicapped child in an institution on a more or less permanent basis. The child may come home for brief periods of time over the summer or on weekends, but will essentially remain in the institution for the rest of his life. Such placement may be necessary because it is not possible to care adequately for the child at home. For example, a bedridden child may become too heavy to move and begin to have problems with skin ulcers and repeated infections. In other instances, behavior becomes totally unmanageable and there is little hope that it can be corrected by a brief placement. Aged parents of handicapped children may not be able to care for their child, or they may die and leave a child who is unable to function independently. In other instances, the physical and emotional effort, time, and money expended on the care of a handicapped child is destroying the family unit. Regardless of cause, the institutionalization of a handicapped child is a painful and difficult process. Permanent institutionalization is generally reserved for children who have profound neurological handicaps that will prevent them from making any adjustment to community life. Individual families vary in their ability to care for a severely handicapped child at home. A social worker can help a family find a placement for their child if it is indicated, and can also help families cope with their feelings about placement. Most families tend to reject institutionalization for as long as they possibly can, sometimes to the detriment of the handicapped child and at other times to the detriment of other children in the family. A social worker may know other families in the area who have gone through a similar experience with whom parents may speak; this may make the process easier.

A social worker will also be able to aid parents in choosing the type of placement that will be best for their child. Most state and private institutions have long waiting lists and cannot take a child for permanent placement immediately. Therefore it is important that parents think about placement early in the course of a child's illness if he has a serious neurological handicap. It may be necessary to wait as long as several years before appropriate placement can be obtained. Making plans well ahead of time will allow parents to be more selective about the institution that their child will enter. A social worker will be able to help a family with the forms that must be completed before a child can be institutionalized; she will also be able to indicate which institutions have long waiting lists. She should also help evaluate a family's financial situation and indicate private institutions that might meet the child's needs better than placement in a state hospital. Parents should visit the different institutions they are considering for their child; they will have to determine which institution, public or private, best meets the child's needs. Parents should inquire about specialized nursing or medical care, what kind of educational or vocational programs the institution offers, and whether the quality of life in such an institution will be an improvement over that which the child would have at home. Parents can discuss these problems with their social worker.

A social worker should be able to inform parents about appropriate observations to make and questions to ask when they visit a facility. Some of these should be: Is the facility accredited or licensed? What is the ratio of staff to children? What therapeutic programs are offered? What are the physical facilities? Is the institution overcrowded? Is it clean? Is the atmosphere stimulating? Is the staff involved with children? What is the quality of daily life? What are the provisions for medical emergencies? What is the reputation of the facility in the community? Is parent involvement encouraged? What are visiting policies? The social worker should help the family become more knowledgeable about specific questions to ask the administrators of an institution, and she should try to locate other parents whose children have been hospitalized at that facility so that the family can talk to them about the institution. A social worker can evaluate a family's situation realistically and discuss the benefits of placement relative to continuing care at home or other alternatives, after the parents have visited the institutions.

It is well to remember that *no* family finds placing a handicapped child in an institution easy. On the other hand, no placement need ever be permanent.

Referrals to appropriate agencies

As we have indicated, children with certain diseases are provided special services by private agencies in the community. Some of these include the Association for Retarded Citizens, the Easter Seal Society, the Epilepsy Foundation, and numerous other organizations that are specifically disease-related. These agencies are funded primarily by contributions from the general public. Some agencies have a paid professional staff and some are staffed by volunteers.

The most valuable aspect of disease-related agencies is that they give parents an opportunity to meet other people in a similar situation. Many parents report that one of the most difficult aspects of raising a handicapped child is the feeling of isolation. Organizations such as the Association for Retarded Citizens can help families meet this need. They can also provide a social structure in which a child can function and his emotional needs can be met. A knowledgeable community social service worker will know which of these associations has interests and programs that are appropriate for a particular child.

In addition to emotional support, an organized group gives families the opportunity to lend a helping hand to someone else. As a group, they can begin to identify specific needs and address them. For example, it might be found that the community lacks babysitters for children who are retarded or who have cerebral palsy. The group may form a babysitting co-op. Some groups have founded schools or established special recreational centers for handicapped children. Other groups have become interested in lobbying for legislative reform. As we indicated in section on finances, this can be extremely important if a child's disability is not covered by a state agency. Even if a family is not in need of state aid, lobbying can be extremely helpful in developing private insurance programs that will aid their child in the future. Social workers frequently advise such groups about internal policies and can direct them to the proper resources for assistance in obtaining and paying for the legal services that are usually needed in any lobbying effort.

REFERENCES

Adams, M. 1960. The mentally subnormal: A social casework approach. Free Press, New York, p. 261.

Beck, H. 1969. Social services to the mentally retarded. Charles C. Thomas, Springfield, Ill., p. 188.

Buscaglia, L. 1973. The disabled and their parents: A counseling challenge. Charles B. Slack, Thorofare, N.J., p. 375.

Heisler, V. 1972. A handicapped child in the family: A guide for parents. Grune and Stratton, New York, p. 156.

PRIVATE AGENCIES CONCERNED WITH NEUROLOGICAL DISABILITIES

Alexander Graham Bell Association
  for the Deaf, Inc.
3417 Volta Place, NW
Washington, D.C. 20007
(202) 337–5220

American Academy for Cerebral
  Palsy and Developmental
  Medicine
1255 New Hampshire Avenue, NW
Washington, D.C. 20036
(202) 659–8251

American Academy on Mental
  Retardation
916 64th Avenue East
Tacoma, Wash. 98424
(206) 442–5462

American Academy of Physical
  Medicine and Rehabilitation
30 N. Michigan Avenue, Suite 922
Chicago, Ill. 60602
(312) 236–9512

American Association for the
  Education of Severely &
  Profoundly Handicapped
1600 W. Armory Way
Seattle, Wash. 98119

American Association on Mental
  Deficiency
5101 Wisconsin Avenue, NW
Washington, D.C. 20016
(202) 686–5400

American Association of Neurological
  Nurses
625 N. Michigan Avenue, Suite 1519
Chicago, Ill. 60611
(312) 944–4280

American Coalition for Citizens with
  Disabilities
1346 Connecticut Avenue, NW
Washington, D.C. 20036
(202) 785–4265

American Council of the Blind
1211 Connecticut Avenue, NW, Suite
  506
Washington, D.C. 20036
(202) 833–1251

American Diabetes Association
600 Fifth Avenue
New York, N.Y. 10020
(212) 541–4310

American Epilepsy Society
Division of Neurosurgery
University of Texas Medical Branch
Galveston, Tex. 77550
(713) 765–3159

American Foundation for the Blind
15 West 16th Street
New York, N.Y. 10011
(212) 620–2000

American Heart Association—Stroke
  Council
7320 Greenville Avenue
Dallas, Tex. 75231
(214) 750–5300

American Occupational Therapy
  Association
6000 Executive Blvd., Suite 200
Rockville, Md. 20852
(301) 770–2200

American Organization for the
  Education of the Hearing
  Impaired
1537 35th Street, NW
Washington, D.C. 20007
(202) 337–5220

American Physical Therapy
  Association
1156 15th Street, NW
Washington, D.C. 20005
(202) 466–2070

American School Health Association
P.O. Box 708
Kent, Ohio 44240
(216) 678–1601

American Speech-Language-Hearing
    Association
10801 Rockville Pike
Rockville, Md. 20852
(301) 897–5700

Arthrogryposis Association
106 Herkimer Street
North Bellmore, N.Y. 11710
(516) 221–6968

Association for Children with
    Learning Disabilities
4156 Library Road
Pittsburgh, Pa. 15234
(412) 341–1515

Better Hearing Institute
1430 K Street, NW, Suite 800
Washington, D.C. 20005
(202) 638–7577

Committee to Combat Huntington's
    Disease
250 West 57th Street
New York, N.Y. 10019
(212) 757–0443

Council for Exceptional Children
1920 Association Drive
Reston, Va. 22091
(703) 620–3660

Deafness Research Foundation
342 Madison Avenue, Suite 1810
New York, N.Y. 10017
(212) 682–3737

Dysautonomia Foundation, Inc.
370 Lexington Avenue, Suite 1504
New York, N.Y. 10017
(212) 889–0300

The Dystonia Foundation, Inc.
425 Broad Hollow Road
Melville, N.Y. 11747
(516) 249–7799

Dystonia Medical Research
    Foundation
9615 Brighton Way, Suite 416
Beverly Hills, Calif. 90210

Epilepsy Foundation of America
1828 L Street, NW, Suite 406
Washington, D.C. 20036
(202) 293–2930

Friedreich's Ataxia Group of
    America, Inc.
P.O. Box 11116
Oakland, Calif. 94611
(415) 655–4964

Hearing, Educational Aid and
    Research Foundation, Inc.
    (HEARS)
1145 17th Street, NW, Suite 942 or
    P.O. Box 57241
Washington, D.C. 20037
(202) 524–5600

Juvenile Diabetes Foundation
23 East 26th Street
New York, N.Y. 10010
(212) 889–7575

Mental Health Association, National
    Headquarters
1800 North Kent Street
Arlington, Va. 22209
(703) 528–6405

Muscular Dystrophy Association
810 Seventh Avenue
New York, N.Y. 10019
(212) 586–0808

Myasthenia Gravis Foundation
15 East 26th Street
New York, N.Y. 10010
(212) 889–8157

National Association of the Deaf
814 Thayer Avenue, Suite 301
Silver Spring, Md. 20910
(301) 587–1788

National Association for Down's
    Syndrome
Box 63
Oak Park, Ill. 60303
(312) 543–6060

National Association for Hearing and
   Speech Action
6110 Executive Blvd., Suite 1000
Rockville, Md. 20852
(301) 897–8682

National Association for Retarded
   Citizens
2709 Avenue E East
P.O. Box 6109
Arlington, Tex. 76011
(817) 261–4961

National Association for Visually
   Handicapped
305 East 24th Street
New York, N.Y. 10010
(212) 889–3141

National Ataxia Foundation
5311 36th Avenue N., Suite 3
Minneapolis, Minn. 55422
(612) 521–2233

National Deaf-Blind Program
Bureau of Education for the
   Handicapped
Room 4046, Donohoe Building
400 Sixth Street, SW
Washington, D.C. 20202

National Easter Seal Society for
   Crippled Children and Adults
2023 West Ogden Avenue
Chicago, Ill. 60612
(312) 243–8400

National Foundation for Jewish
   Genetic Diseases
609 Fifth Avenue, Suite 1200
New York, N.Y. 10017
(212) 753–5155

National Foundation—March of
   Dimes
1275 Mamaroneck Avenue
White Plains, N.Y. 10605
(914) 428–7100

National Genetics Foundation, Inc.
9 West 57th Street
New York, N.Y. 10019
(212) 759–4432

National Hearing Association
1010 Jorie Boulevard, Suite 308
Oak Brook, Ill. 60521
(312) 323–7200

National Hemophilia Foundation
25 West 39th Street
New York, N.Y. 10018
(212) 869–9740

National Neurofibromatosis
   Foundation
340 East 80th Street, #2111
New York, N.Y. 10021
(212) 744–4601

National Rehabilitation Association
1522 K Street, NW, Suite 1120
Washington, D.C. 20005
(202) 659–2430

National Reyes Syndrome
   Foundation
509 Rosemont
Bryan, Ohio 43506
(419) 636–2679

National Society for Autistic Children
1234 Massachusetts Avenue, NW,
   Suite 1017
Washington, D.C. 20005
(202) 783–0125

National Spinal Cord Injury
   Foundation
369 Elliot Street
Newton Upper Falls, Mass. 02164
(617) 964–0521

National Tay-Sachs and Allied
   Diseases Assn., Inc.
122 East 42nd Street
New York, N.Y. 10017
(212) 661–2780

National Tuberous Sclerosis Assn.,
   Inc. (NTSA)
P.O. Box 159
Laguna Beach, Calif. 92652
(714) 494–8900

The Orton Society, Inc.
8415 Bellona Lane, Suite 113
Towson, Md. 21204
(301) 296–0232

Rehabilitation International USA
20 West 40th Street
New York, N.Y. 10018
(212) 869–9907

Spina Bifida Association of America
343 S. Dearborn, Room 319
Chicago, Ill. 60604
(312) 663–1562

Tourette Syndrome Association, Inc.
42–40 Bell Boulevard
Bayside, N.Y. 11361
(212) 224–2999

Tuberous Sclerosis Association of
    America
P.O. Box 44
Rockland, Mass. 02370
(617) 878–5528

United Cerebral Palsy Association,
    Inc.
66 East 34th Street
New York, N.Y. 10016
(212) 481–6300

11

THE NURSE

Nurses are traditionally thought of as hospital-based professionals caring for acutely ill children and adults, as members of a clinic staff who assist physicians, or as assistants in the doctor's office. However, the role of the nurse in the care of children is changing dramatically. The spiraling costs of hospitalization have led to increased emphasis on ambulatory care. Nurses are now better prepared to assume more complex roles in the delivery of health services. Many specialize in a single area of care after their initial training. The services of these nurse specialists have become increasingly available to the chronically handicapped child and his family.

Parents who seek nursing services should first identify the specific needs of their child. This might best be accomplished in consultation with the child's physician, since he would be familiar with both the child's medical needs and any special needs within the family. Having assessed the kinds of expertise needed, the physician should be able to direct the family to the appropriate agency for nursing services.

Nurses vary in their basic educational training. Training programs last from two to four years. All graduates then take the same licensing exam and receive the same title, Registered Nurse (R.N.). They may then choose to prepare for a master's degree in a special branch of nursing, such as pediatrics, with a primary focus on handicapped children. These nurses are given advanced training in the effects of different handicapping conditions on growth and development, nutrition, and feeding. In addition, they learn to work with other professionals in the community to secure needed services for the children and their families.

Other nurses take a six- to twelve-month program leading to a special certificate as a Pediatric Nurse Practitioner. These nurses are taught to do physical examinations, developmental evaluations, and to manage common medical problems under the supervision of a physician.

It is not the purpose of this chapter to discuss the usual duties of the floor or clinic nurse who helps care for a handicapped child who becomes acutely ill. The section that follows focuses on those nurses who help families provide continuing care for their handicapped children.

THE PUBLIC HEALTH NURSE

The public health nurse generally does not have an advanced degree. However, she is often the person who is most available to a family at their home. Therefore, she is most able to meet a family's ongoing need for nursing care. She collaborates with physicians, social workers, teachers, physical and occupational therapists, and others involved in a child's medical care. In that capacity, she is able to help parents understand their child's problem more fully, and she can answer questions that arise as a long-term treatment plan is put into effect at home. She can help parents judge the effects of medication on their child, evaluate the child's behavior, and often give parents advice about managing minor behavioral problems or, if necessary, direct them to an appropriate facility. It is not unusual for the public health nurse to provide information about the management of minor health problems and to assist parents in providing for more complex needs, such as tube feeding, suctioning, and bowel or bladder care. A public health nurse is knowledgeable about local agencies and services that can be mobilized to assist the families of handicapped children.

Some public health nurses are trained to assist parents to develop and/or carry out programs designed to facilitate developmental progress. This type of public health nurse is equipped to guide parents who are teaching their child self-care skills, such as toileting, feeding, and dressing, and she can indicate when the child is ready to be taught these skills. More important, because of her close association with the family, she is often in a position to help them select an appropriate program for their child. Before parents accept the advice of a public health nurse on matters of behavior and development, they should feel free to inquire into her training in these areas. The public health nurse can be contacted by calling the county health department.

In localities where there are both public health and visiting nurse services, each group may have a different set of responsibilities. For example, the visiting nurse might only provide direct nursing care services at home, while the public health nurse may restrict her activities to the teaching of health care. The nurse associated with the local office of the state's Crippled Children's Services can assist parents in obtaining services and procuring needed equipment and supplies for their child. This nurse is usually available at the clinics operated by Crippled Children's Services. She can answer questions parents might have about their eligibility for Crippled Children's programs and assist them with their application for help.

THE SCHOOL NURSE

School nurses can help families of handicapped children in a variety of ways. If a child has epilepsy, for example, she can be helpful in giving medication and in monitoring the child's seizure activity and behavior as various changes are being made in his drugs or their dosage. In most states, if a child is taking medication, a school nurse must have written instructions from a physician and written permission from the parents to give the child medication during school hours. The school nurse can also help educate the child's teachers and his fellow students about epilepsy or other handicapping conditions, thus minimizing fears or discrimination on the part of the staff or the child's classmates. Teachers and classmates who understand a child's disorder are more likely to make positive contributions to that child's school experience and encourage good social-emotional adjustment.

On occasion, the nurse can help children with more specialized needs return to the school system. There are children at public school who have tracheostomy tubes, for example, that require periodic suctioning. Other children may need to be tube-fed in order to stay in school. The school nurse can provide these services.

School nurses also are active in providing immunizations and teaching preventive health care measures, including dental care, personal hygiene, and sex education. How active they are in any of these areas depends on the local school board, the school's principal, and the interests of the particular school nurse. Her work responsibilities may restrict her teaching activities. She is also less apt to be involved in providing certain services for a handicapped child if there are experts in other disciplines, such as social workers or occupational and physical therapists, affiliated with the school. However, even under these circumstances, the school nurse is sometimes involved in the conferences that are held to design a child's Individual Educational Program to make sure that health concerns are part of the overall plan.

Parents can request that the school nurse be available to their child during school hours for any of the activities described above. If she cannot be made available, parents may appeal to the school board. If the lack of suitable nursing care is a major reason why a child cannot be placed in a public school, the Board is obligated to propose an alternative method of education.

THE PEDIATRIC NURSE PRACTITIONER (PNP)

These are nurses who have completed a special program to gain additional knowledge and skills in providing primary health care to pediatric patients. They do physical examinations and treat common problems such as colds, ear infections, diarrhea, and skin lesions. They work under the direction of a physician in an office or clinic, and frequently assume responsibility for supervising the general health of the children who visit that office or clinic. Depending on the particular setting in which the pediatric nurse practitioner works, she may be able to assist families in many ways that are similar to the public health nurse, in addition to providing this general health care. A PNP cannot sign prescriptions, and any specific therapeutic program she plans for a child must be done in cooperation with a physician. The PNP and MD work as a team. Parents should know which doctor is responsible for the care rendered their child by a PNP in any clinic or group setting.

THE CLINICAL NURSE SPECIALIST

The clinical nurse specialist, whose area of expertise is in developmental disabilities, is a nurse with a master's degree in nursing with specialized training, knowledge, and skills in the area of the developmentally disabled child and adult. Often nurses with this type of preparation are associated with an interdisciplinary team affiliated with a medical school, large hospital, or clinic. The team is designed to provide a comprehensive evaluation of a client's and family's need for services, generally on an outpatient basis, and then to assist the client and the family in obtaining needed services in the community, while giving limited treatment themselves.

The clinical nurse specialist on a team of this type may act as a case manager. In this capacity, she assists in coordinating the services that the child and his family will receive. She helps parents locate educational and treatment facilities in their community. She is also available to answer questions and to assist the child and his family in dealing with various problems as they arise. She collaborates or works with other professionals on the team in developing and implementing programs that facilitate developmental progress, in managing feeding and behavior problems, and in helping the child develop self-help skills.

Clinical nurse specialists with master's degrees in developmental disabilities, maternal-child health, or pediatric nursing can also be found in inpatient units of major medical centers where handicapped children are served. How these nurses function and what services they provide depends on their position within the hospital organization.

The clinical nurse specialist may be found on the patient care unit where a handicapped child is admitted for evaluation. She is usually employed by the hospital and will evaluate a child at the request of the child's family or his physician. In addition to the activities already mentioned, she is available to explain certain aspects of the child's illness and its management. Sometimes these may not be clear to the family after an initial discussion with the child's physicians, or questions may arise later as a treatment plan is implemented. Moreover, because of her availability on the patient care unit, she may have more time to spend with families, and therefore can act as a valuable liaison between physician and family. If necessary, she will, in cooperation with the regular staff floor nurses, help teach the family the nursing skills they will need to care for the child at home.

After a child is discharged from the hospital, the clinical nurse specialist may see him and his family on a regular basis when he returns for outpatient visits. In some settings, the clinical nurse specialist may also have pediatric nurse practitioner skills and will examine the handicapped child and talk to his family with the physician at the time of follow-up visits. Often, she will keep in phone contact with the family about medical problems and, in some instances, if no visiting nurse is available, she will make home visits to supervise and help the family improve their nursing skills.

The clinical nurse specialist is available to talk with other professionals in schools, developmental centers, clinics, and agencies regarding specific patient problems. As the need for services is identified, the nurse can help parents locate the appropriate community resources or refer them to other professionals or agencies for assistance. She is also available to talk to various parent groups and organizations regarding handicapping conditions and their management.

When a number of professionals are involved in providing services to a handicapped child and his family, differences of opinion about the condition of the child and the nature of his disease or its prognosis may occur. In these situations, where medical professionals of various disciplines work as a team, they frequently confer. There should be little difference in the information and opinions provided the family of a handicapped child. However, when such differences of opinion do occur, they should be discussed and clarified with the physician, together with the other professionals involved. It is the physician who is most familiar with the child's medical problems and who is ultimately responsible for the child's medical care.

In summary, it can be seen that the functions of nursing specialists, nurse practitioners, school nurses, and public health nurses often overlap with each other and with other medical professionals. Nurses are available to families in a variety of capacities, depending on the agencies with which they are associated and their educational preparation. Therefore, in seeking nursing services, it is important to find out what help can be expected from a nurse in a particular situation.

REFERENCES

Bumbalo, J. A. The Clinical Nurse Specialist. *in* K. E. Allen, V. A. Holm, and R. L. Schiefelbusch, eds. Early intervention—A team approach. University Park Press, Baltimore, pp. 123–145.

Cook, C. L. Nursing Services. *in* R. Koch and J. C. Bobson, eds. The mentally retarded child and his family. Brunner-Mazel, New York, pp. 206–218.

Guidelines for continuing education in developmental disabilities. 1978. American Nurses Association, Kansas City, Mo., Publ. code NP–58.

12

THE LAWYER

Parents of handicapped children need to be better informed about the laws that secure and protect their own and their child's basic rights. Over the past twenty years parents, concerned educators, and other advocates have worked together to secure official recognition of the human and civil rights of all handicapped children and adults. Progress has been made at all levels and in all branches of government to assure, at least in policy, the right of handicapped persons to participate in a broad spectrum of life activities. New strong federal laws have been enacted that reflect this official recognition. However, if the new philosophy and the laws that flow from it, which accept each person's right to learn to the extent of his or her ability, to become as independent as possible, and to take an active part in society, are to be translated into reality, parents must know, understand, and use their rights.

The purpose of this chapter is to acquaint parents of handicapped children with some important legal issues that have relevance for them and their child, and to provide these families with information about pertinent state and federal laws and regulations. The material presented here *is in no way intended to be taken as legal advice* nor is it meant to be an in-depth summary of the entire body of law pertaining to handicapped American children or adults. We focus on four issues that have a major impact on the lives of handicapped children and their parents:

1. Legislation and legal factors relevant to the handicapped child within the educational setting.

2. Legislation and legal factors relevant to the handicapped person's vocational training, employment, and full participation in community services.

3. Legislation governing parental obligation to support a handicapped child and guardianship proceedings.

4. Financial assistance provided by law to handicapped children.

The rights of all citizens are protected by a clause in the Fourteenth Amendment to the United States Constitution that guarantees to all people equal protection under the laws. However, interpretation and implementation of the equal protection clause as it applies to handicapped persons varies from state to state and in some instances from agency to agency within a state. It would be impossible to cover a subject matter so large in scope and of such a volatile nature (i.e., subject to continued court interpretation and state regulatory changes) in a single chapter. Therefore, parents are reminded that the content of this chapter is meant to serve only as a guide or a general index on the subject matter contained. The information presented here is drawn mainly from the author's construction of federal and current Missouri law on the subject. Reasonable legal minds could vary and disagree with the author's conclusions. Parents who have questions about the laws and regulations in their state should seek appropriate legal assistance to ensure protection of their child's interests.

EDUCATION—LEGISLATION AND LEGAL FACTORS

The school district's responsibilities

Traditionally the American public has accepted the proposition that a public education is due as a matter of right to every American child, that is, every child except the handicapped child. Before 1977, millions of handicapped children were denied any education while others were given token educational services mainly in the belief that a public school did not have to provide equal facilities for handicapped children. The Education For All Handicapped Children Act (P.L. 94–142) was passed by Congress in 1975 and went into effect in September 1978. This law establishes the right of *all* American children to an education. In the exact words of the law, its purpose is: "to assure that all handicapped children have available to them . . . a free and appropriate public education which emphasizes special education and related services designed to meet their unique needs, to ensure that the rights of handicapped children and their parents or guardians are protected, to assist states and localities to provide for the education of all handicapped children and to assess and assure the effectiveness of efforts to educate handicapped children." This landmark legislation makes it possible for states and localities to receive federal funds to assist in the education of all handicapped children. The definition of "handicapped" as set out under the law is very broad. Children are considered handicapped if they are found to be mentally retarded, speech impaired, visually handicapped, emotionally disturbed, orthopedically impaired, deaf, blind, multi-handicapped, handicapped by other health impairments, or suffering from specific learning disabilities. To qualify for special education, children must not only have a handicapping condition; they also must be found in need of special teaching and/or other related services because of their impairment. P.L. 94–142 covers children from age three to age twenty-one, unless state law does not provide for public school services for children three to five or eighteen to twenty-one.

Children with special learning needs are therefore entitled to free special services to meet their needs. The primary responsibility to ensure this education is given to the local school district; thus, it is important for parents to know the pertinent laws of their state and locality. The ages of children eligible for public education vary in different states. Some states provide services for handicapped children at birth with infant education programs and other help for parents. Other states extend the age of eligibility past twenty-one. Each state has exact procedures for receiving services; parents should secure a copy of their state's special education law and the recent amendments or attorney general's decisions related to the education of handicapped children. To receive federal funds for special education under P.L. 94–142, every state *must* submit plans to the federal government *each* year to show how they intend to provide a free and appropriate education for all handicapped children. *These plans must be made available to the public for review and comment before they are adopted and sent to Washington.* Dates for review must be announced far enough in advance for parents and other interested persons to appear at hearings and to express their views.

P.L. 94–142 also makes it possible for local school systems to receive a share of federal funds as long as the requirements specified by law are met. Local school districts may not provide less than the state law requires, but they may provide more. Thus the services provided by local school districts within a state may not be identical. Local school districts must prepare detailed plans for state approval demonstrating how they will fulfill the purposes of the law. Parents should contact their state or local director of special education to find out when hearings will be held and to request copies of plans that are to be submitted to either the state or to the federal government for approval.

It is very important for all parents of handicapped children to know the key provisions of P.L. 94–142:

1. First priority must be given to handicapped children who are not now receiving an education and second to the most severely handicapped children whose education is inadequate.

2. The rights of parents and children must be guaranteed by due process by states and localities. The parents' rights must be protected in all procedures related to the identification, evaluation, and placement of their child and the provision of an appropriate education for him or her. Procedures must also be provided to afford the opportunity to protest educational decisions made by school officials.

3. Children may be placed in special or separate classes only when it is impossible to work out a satisfactory placement in a regular class with supplementary aids and services.

4. All testing and evaluation must be racially and culturally nondiscriminatory and must be in the child's primary language or mode of communication. No one test or procedure may be used as the sole means of making a decision about an educational program.

5. Parents may participate on the team that draws up the individualized educational plan for their child. If it is appropriate, the child may be included in this planning program. The law is very clear about the prescriptions for the inclusion of short- and long-term educational goals and the specific services that are to be provided. Since 1977, the law has required that individual plans be reviewed at least annually and revised according to the child's changing needs.

6. Children who are placed in private schools by state or local education systems in order to receive an appropriate education must be given this education at no cost to parents or guardians. The private school programs must meet standards set by law, and such schools must protect the rights of parents and children as guaranteed by law.

7. Each state must set up an advisory panel including handicapped individuals and teachers and parents of handicapped children. This panel is meant to advise the state on unmet needs, comment publicly on rules and regulations, and assist in evaluating programs.

8. Funds can be withheld if, after reasonable notice and opportunity for a hearing, the state is found by the U.S. Commissioner of Education to have failed to comply with the provisions of the Act.

9. Payments by the state to local school systems may also be suspended for noncompliance. If the state determines that a locality is unable or unwilling to set up or consolidate programs or has children who can best be served by regional or statewide programs, the state will use the funds to provide services directly to those children.

Due process—equal protection under the law

Due process is a constitutional right, and it protects the liberty of all citizens. It means that an individual may not be deprived of his or her rights as a free citizen unless very specific procedures guaranteed by law are followed, including a fair chance for the individual's side to be heard by an impartial person, such as a hearing officer, a jury, or a judge. The Education For All Handicapped Children Act has a strong provision mandating due process in all matters relating to decisions about special education. To qualify for funding under the new law, states *must* guarantee certain procedural safeguards. If these procedures are not written into the education law in a state, they can be found in the plan that the state must submit to receive its share of federal funds for education of the handicapped. It is essential for every parent to know what these due process rights are. This is the one way in which they can be sure that the law really works. It is their basic means of making sure that their child gets the education he needs. The Education For All Handicapped Children Act protects the parent or guardian's right to be informed of and to participate in all the important decisions made about a child's schooling. Parents are entitled to:

1. Notice in writing before any action is taken or recommended that may change or fail to change a child's school program.

2. All records related to the identification, evaluation, or placement of a child.

3. A due process hearing when parents and the school cannot agree on what type of program will be most helpful for a child. The law specifies rights available to all parents at such hearings to ensure fairness and impartiality.

4. The right to appeal if parents are not satisfied with the results of a due process hearing. The appeal is made to the State Department of Special Education. If disagreements remain after all of these procedures have been exhausted, parents have the right under law to appeal to the courts. Most states now provide a written rights guide that puts all the steps parents must take to obtain special education services for their child into clear, simple, nontechnical language.

Evaluation by school districts

P.L. 94–142 requires the consent of parents before any school district can determine which special education services are necessary or transfer a child from a regular education program to a special education program. The parents, a teacher, a school administrator, or any person first suspecting that a child is in need of special education can request an evaluation be made. If the school recommends this evaluation, parents must be notified in writing of the purpose in requesting the evaluation, the type of tests to be given, and any reasons for recommending a change in the child's existing program. The evaluation is generally made by the Department of Special Education within the local school district or by a person qualified in the administration and interpretation of psychological tests who can determine from the test results what special education services are necessary.

Appeal from the school district's findings

Built into the law are protections to ensure a fair, careful, and accurate evaluation of the special needs of the child. If after the evaluation has been completed parents are not satisfied with the results or the suggested educational service to be given the child, the Act gives them the right to challenge the school's findings if they believe any of the following conditions exist:

1. The evaluation was inadequate.
2. A fair representation of the parent's thoughts and ideas was not included in the evaluation.
3. The placement recommended for the child is not appropriate for their child's individual abilities and needs.
4. The child is showing no progress or insufficient progress in the educational program the school is providing.

The Act sets out in detail the grounds on which the local school district's findings may be appealed. If all lines of communication, discussion, and settlement are terminated, the Act sets out five courses of action parents may pursue: (1) an independent evaluation; (2) a hearing before a neutral hearing officer; (3) an administrative appeal; (4) a complaint to the federal office for civil rights; and (5) a lawsuit.

Family Educational Rights and Privacy Act of 1974

This federal act applies to all children, not just those who are handicapped. The Family Educational Rights and Privacy Act of 1974 spells out and clarifies the rights of students and parents to examine records kept in the student's personal files. Under this legislation, parents have the right to examine these records if they so request. In the event of litigation, parents can obtain copies of all records, the theory being that parents can assist the teacher in making the educational decisions for their children. This Act in part gives parents the right to inspect and review their children's educational records, have these records explained, and to the extent that these records contain inaccurate or misleading information, to have them amended.

VOCATIONAL TRAINING, EMPLOYMENT, AND PARTICIPATION IN COMMUNITY SERVICES— LEGISLATION AND LEGAL FACTORS

Vocational Educational Amendment

A law as important as P.L. 94–142 is the Vocational Education Amendment of 1976 (P.L. 94–482). This act was passed to strengthen the abilities of states to provide vocational education, especially to those groups whose need is most acute. This law has an important relationship to P.L. 94–142 and can contribute tremendously to a child's future employability. The purpose of the law is to extend help not only to handicapped children but to disabled individuals of all ages who need vocational education. The law requires that:

1. Ten percent of federal funds allocated to states for vocational education be spent for special programs, services, and activities for the handicapped. Special funds may go to school districts for secondary school programs and also to post-secondary adult vocational education programs and community colleges.

2. Local education systems and post-secondary schools must apply to states for a share of this money. Funds are not allocated automatically.

3. Programs using these funds for vocational education in secondary schools must carry out the goals of P.L. 94–142 and comply with its requirements. All of the rights guaranteed under P.L. 94–142 are extended to handicapped children served under the Vocational Education Act. All rights to due process are protected, including the right to participate in decision making, to accept or reject proposed education plans, and to ask for a hearing and appeal if parents are dissatisfied with the programs offered.

4. Vocational education plans for an individual child must be coordinated with the child's IEP.

Advisory councils must be set up on both state and local levels and have an equal representation of consumers [parents] and professionals [vendors] who provide services. The law contains other requirements, such as that vocational services for handicapped children include vocational instruction, curriculum development and modification to enable handicapped students to take part in regular programs, modification of vocational equipment to enable students to develop skills leading to employment, vocational or work evaluation, supportive services (e.g., interpreters for deaf students, note takers, readers, tutorial aids, vocational guidance and counseling, job placement, and follow-up services). Parents can get copies of proposed plans and information about when and where hearings will be held and other facts about vocational education in their area by contacting appropriate state and local officials. Each state has a coordinator of vocational education for the handicapped in the Department of Education under the Division of Vocational Education. Parents can find out through their district superintendent of schools who is responsible for the vocational services in their local school system.

Employment practices in hiring handicapped persons

Definition of handicapped person   Section 504 of the Federal Rehabilitation Act of 1973 covers all rights of handicapped people of all ages in all areas of life. It is a civil rights law that prohibits discrimination on the basis of handicap in any private or public program receiving federal financial assistance. Section 504 also contains the general provisions for employment of handicapped persons. In general, subpart B of this Act deals with employment practices (i.e., discrimination against a qualified handicapped person). The Act defines a qualified handicapped person for the purposes of seeking employment as "one who, with reasonable accommodation, can perform the essential functions of the job in question."

Job discrimination on the basis of handicap   If a handicapped person seeks employment, it must first be determined what the job requirements are and next whether the handicapped person applying for the position can be accommodated to perform the duties. Even though it may be necessary to modify the method for the performance of the job in order to employ the handicapped person, the test is "reasonable accommodation to the known physical or mental limitations of an otherwise qualified employee." However, if the accommodation imposes an undue hardship on the employer, employment of the handicapped person may not be required. There may be differences of opinion with regard to the degree of accommodation needed and the meaning of the term "undue hardship" as it applies to the employer. If there is disagreement, the person who believes that he has been discriminated against should file a written complaint with the State Regional Office for Civil Rights, Department of Health and Human Services. The complaint must be made within 180 days of the violation and should contain the complainant's name, address, and telephone number as well as a description of the violation. A statement of the complainant's handicapping condition should also be included. The Office for Civil Rights will conduct an informal investigation and will notify the complaining person, in writing, if it concludes that no discrimination occurred. If the investigation indicates that a violation of Section 504 has occurred, the Office will try to resolve the matter through informal means. However, if no remedy is forthcoming to correct the violation, the Office for Civil Rights has the power to suspend or terminate all federal funds to the organization, refer the case to the Department of Justice, or refer the case to the appropriate state authorities for enforcement. Section 504 includes provisions that clearly protect complainants or others who have participated in an investigation of employment discrimination from retaliation by an employer.

Participation in community services

Handicapped persons must have opportunities to participate in or benefit from social and recreational programs equal to those that are provided to others. Included in Section 504 is a provision which states that "All programs and services must be barrier-free. Programs must be made accessible immediately by means of changing locations or providing alternative ways for handicapped persons to use facilities and equipment."

PARENTAL OBLIGATION TO SUPPORT A HANDICAPPED CHILD AND GUARDIANSHIP PROCEEDINGS

Parental obligation

Sometimes, after a parent of a handicapped child has accepted the fact that his or her child will need lifelong medical attention, supervision, and care, the question of providing support for this child arises. In a stable marriage, both husband and wife understand and assume the financial responsibility for continued care of the handicapped child, irrespective of his age. Most of these parents budget for this care from within and from outside the primary family unit. It is primarily in those families in which the parents have separated or are in the process of seeking a divorce that both mother and father want a definitive restatement as to the legal liability of the father for continued support of the handicapped child once he reaches the age of twenty-one. It is generally assumed, and legally correctly so, that the natural father of the handicapped child will be responsible for all the care, support, and medical attention the child needs up to the age of twenty-one. Of course, the amount of additional support needed because of the child's handicap is weighed along with the financial ability of the father to pay.

There are two questions commonly asked an attorney with regard to the legal responsibilities of separated or divorced parents. The first is asked by the father: Is there a responsibility or legal obligation to support my unmarried, solely dependent child who is over the age of twenty-one? The second is asked by the mother: What is the proper legal procedure to follow to obtain support for my handicapped child who is either (1) approaching the age of twenty-one; or (2) over twenty-one but no support has been received since the child reached twenty-one.

In response to the father: As a general proposition of the law, a parent's obligation to support a child ceases when the child marries, reaches his majority (twenty-one), or leaves the parent's residence in order to emancipate himself. This varies from state to state. Missouri case law, for example, clearly sets forth the liability of a parent to support an unmarried, unemancipated, solely dependent child over the age of twenty-one:

> "A parent has a duty to support his child who is unemancipated, unmarried, and solely dependent past the age of 21" (*State of Missouri* vs. *Honorable Michael J. Carroll*, 309 S.W.2nd. 654). "The obligation of a parent to support a helpless adult-child terminates when the necessity of support for said child ceases" (60 C.J.S. Parent and Child Section 17, p. 704).

Once a court determines that the handicapped individual (adult-child) is incapable of supporting himself by virtue of his handicap, the primary obligation of the father is to provide financial assistance for this adult-child even though he has attained the age of twenty-one. This obligation can terminate in part or in whole at such time as the disability is removed and the adult-child becomes either partially or completely self-supporting.

The separated or divorced mother's question usually arises in one of two situations. The first is that the parents of the handicapped child were divorced before the child attained the age of twenty-one, but the child is now so close to twenty-one that the immediate concern is "what to do now." Second, the mother, who has custody of the child, is told that no further child-support will be provided since the adult-child has attained the age of twenty-one. The divorced mother tries to explain the law or her interpretation of the law to the father, but he still fails to provide support. Shortly thereafter, the mother appears in a lawyer's office, explaining the absolute necessity for additional support for the handicapped adult-child.

The first issue to be resolved in this conflict is, could the divorce judge have ordered continued support for the child after the age of twenty-one, if the Court were aware of the mental or physical condition of the child during the divorce and before the child attained the age of twenty-one? The answer (in Missouri) is no. The power of the divorce court to make orders touching the support of a child ceases when he attains the age of twenty-one (452.070, R.S.MO. 1969). When the child-adult attains the age of twenty-one, he may proceed in his own right to establish the obligation of support from his father (*Martin* vs. *Martin*, 539 S.W.2nd. 756). Again, other states may differ, and legal advice is needed to establish the case law in the state in which the child resides.

If the child is a minor, suit must generally be brought in the jurisdiction where custody was originally given to the parent who has the child. The handicapped adult-child who has a legal disability will need the services of a legal guardian (usually the mother) to bring the suit for support in his name. This legal guardianship is discussed in more detail below.

Guardianship proceedings to handle monies of handicapped children

Appointment of the parent as guardian   Section 475 VAMS grants to the father and mother, with equal powers, the rights and duties over their minor child (under age twenty-one) while they are living, and in the event of the death of either parent, to the survivor. The most common time a lawyer is consulted, concerning the necessity of having a guardianship estate established for the benefit of a handicapped child, is when a relative decides that he or she wants to "leave" the child some money or other valuable in a will and wants to know if any "peculiar" legal problems will arise in order for the handicapped child to receive the money.

Any child, up to the age of eighteen, according to Missouri law (and that of many other states), does not have the legal capacity to receive bequests by will. A child over eighteen who is handicapped (at least as to his mental ability) will not have the mental capacity to receive bequests by will either. A vehicle must be created to accept the money and hold it for the use, enjoyment, and benefit of the handicapped child. This vehicle is called a guardianship estate. There are essentially two different types of guardianship estates:

1. Guardianship of the estate of the handicapped child can be created in the probate court in any instance where the person is adjudged incompetent. Where the minor has no parent living or where there is a natural guardian (parent) but the court finds that the best interest of the minor requires guardianship, a guardianship of the minor's estate can be created. This is for the purpose of administering any assets owned by the minor or to be acquired by the minor through bequest or other gift.

2. Guardianship of the minor may be granted either when the minor child has no parent living or where if living the parent is adjudged unsuitable or incompetent for the duties of guardianship or where the father of a minor is imprisoned in the penitentiary of the state.

It is to be noted that no guardian is appointed for a married minor.

Guardianship of a handicapped child, residency in a state residential facility care center Once a child attains the age of eighteen (in Missouri, other states may differ), unless he or she has been declared "an incompetent" by a court of jurisdiction in the county where he or she resides, the residential facility care center will not be able to hold (i.e., provide residence and security) the child against its will (Missouri House Bill 561).

This means that unless an incompetency proceeding is instituted, and a "guardian of the person" appointment is made by the court, at age eighteen the handicapped child may no longer be confined to a residential facility care center without his permission.

FINANCIAL ASSISTANCE AVAILABLE TO A HANDICAPPED CHILD

Social Security disability benefits to handicapped children

Because of the variety of benefits supplied by Social Security, this section will deal only with the Supplemental Security Income Program (SSI). This federal probram makes monthly cash payments to disabled persons with limited assets in their own name, as well as limited income. Age is no criterion, i.e., a disabled child as well as a disabled adult can receive SSI payments.

Definition of disability For Social Security purposes, disability means the inability to work because of a severe physical or mental impairment that has lasted (or is expected to last) at least twelve months or to result in death or blindness (either central visual acuity of 20/200 or less in the better eye with the use of corrective lenses or visual field reduction to 20 degrees or less) (U.S. Dept. of Health, Education and Welfare Publication No. SSA 78-10002, May 1978).

To qualify for SSI, "a retarded child under eighteen (or twenty-one and attending school) is considered disabled if it is shown that:

1. I.Q. is 59 or less or

2. I.Q. is 60 through 69 with marked dependence upon others for meeting basic personal needs and a physical or other mental impairment that restricts function and development or

3. The achievement of developmental milestones is not greater than that generally expected of a child half the applicant's age" (DHEW Publication No. SSA 78-11050, May 1978).

For an adult retarded person to qualify for SSI, his or her own income and assets are considered in determining eligibility. A retarded adult is considered disabled if it is shown that:

1. I.Q. is 49 or less.

2. I.Q. is 50 to 69 with an inability to perform routine, repetitive tasks, and a physical or other mental impairment resulting in restriction of function.

3. There is severe mental and social incapacity which creates needs, characterized by inability to understand the spoken word, inability to avoid physical danger or follow simple directions, and inability to read, write, and perform simple calculations (U.S. Department of Health, Education and Welfare Publication No. SSA 78-11050 May, 1978).

Personal assets and eligibility for SSI An individual can be eligible for SSI if he or she has assets with a value of $1,500 or less. As a general rule, such personal assets as household goods, wearing apparel, insurance policies, and an automobile do not count in determining eligibility. The main assets used in determining eligibility are savings and checking accounts, cash, savings certificates, stocks, bonds, and other securities.

The mere fact that a person has income will not restrict his right to receive SSI. As a general rule, the first $20.00 a month income from such sources as bonds, stocks, bank accounts, interest, and generally those sources that would be listed on an income tax return under

Schedule B Form 1040 are excluded in determining qualification for benefits.

State financial assistance available to handicapped children

Application for acceptance by a Department of Mental Health facility
Each state varies in its application requirements, but the general information sought is the condition of the applicant relative to his/her prenatal, birth, and developmental history; medical information, and educational history. Attached to the application are medical authorization forms allowing state investigators to obtain medical information on the condition of the applicant. In Missouri, for example, direct financial assistance is available for a handicapped child through a state contract for funds administered through the Department of Mental Health. The contract is a year-to-year agreement allowing the parent or guardian to receive state financial assistance depending on four main factors:

1. Income of parent or legal guardian.
2. Income of handicapped child.
3. Actual expenses incurred by a parent or legal guardian in having a child reside in a state-approved facility (room, board, clothing, medical and personal allowance).
4. Department of Mental Health determination as to what is reasonable cost of care of providing room, board, clothing, medical and personal allowance.

To apply for a Missouri Department of Mental Health Contract, a parent first completes an "Applicant Questionnaire" and mails it to the Regional Department of Mental Health Center in the district in which the child resides. Residents of states other than Missouri should contact their state Department of Mental Health for application forms and instructions.

Title XX Funds Title XX Funds are federal funds administered through the states and in particular through the individual state's Department of Mental Health. The purpose of Federal Title XX Funds available to states is to encourage each state

1. to help achieve or maintain economic self-support for its handicapped children.
2. to achieve or maintain self-sufficiency, including reduction or prevention of dependency of its handicapped children.
3. to prevent or remedy the neglect of children unable to protect their own interests.

4. to reduce institutional care by encouraging community based care (i.e., care away from state institutions)

5. to secure referral or admission for institutional care when other forms of care are not appropriate as set out under 4 above (Sec. 2001 Title XX, effective October 1, 1975).

The Act allows direct payments to states in such amounts as are appropriated by the federal government and applied for by the individual states. Limitations are placed on states as a qualifying condition in obtaining these funds, such as the income of the recipient (or the gross income of the family of which the individual is a member) not being in excess of the median income as determined by the federal government.

The Act enumerates in detail fifteen restrictions that limit a state's right to receive these funds. No payment may be made under this Act to any state with respect to any expenditure in part for:

1. Medical care, except for the initial detoxification of an alcoholic or drug dependent individual for a period not to exceed seven days.

2. Purchase, construction on or major modification of any land, building or other facility.

3. For goods or services provided by a private institution.

4. An expenditure which is made from donated private funds.

5. Room or board (Sec. 200 (a) 7-A).

States that accept the benefits of the Act are required to adhere to a strict reporting system about their use of the funds (Sec. 2003 (a)). The federal government is given the discretion to suspend further payments to the states if a finding is made that there has been substantial noncompliance with the Act, but procedures are available within the Act for a fair hearing by the state before a state agency is denied further assistance (2001 (a) (1) A).

Each state is given the right to select its own service program, defined in the Act as Services Program Planning (2004 A), and a time limit is set forth for submitting the program. The Act finally provides for the evaluation of the state programs and for assistance, when requested by the states toward the implementation of these programs (2006 (a)).

It is the primary responsibility of a social worker to inform parents of existing programs, but all questions concerning the legal rights of a handicapped child should be referred to a lawyer.

SUGGESTED ADDITIONAL READING MATERIAL

Children's Defense Fund Publication: *A Guide For Parents And Advocates,* 1520 New Hampshire Ave., N.W., Washington, D.C. 20036

Missouri Digest, Schools, Sec. 154

Child Welfare League of America, Inc. Publication: *Information on Sec. 504: Non-Discrimination on the Basis of Handicap,* dated April 23, 1979, 67 Irving Place, New York, N.Y. 10003

Social Security Benefits for People Disabled Before Age 22, HEW Publication: *No. (SSA) 78–10012*

Applicant Questionnaire, Application for Admission to a Missouri Division of Mental Health Facility, Marshall Regional Center, P.O. Box 190, Marshall, Mo. 65340

Explanation of Title XX Funds Publication: Mental Affairs Office of the National Association for Retarded Children, Vol. 3 No. 7, May 22, 1975, 1522 K Street, N.W., Washington, D.C. 20005

Federal Register Publication: *Monday, January 31, 1977, Part III, Title 45, Part 228, Social Securities Programs for Individuals and Families: Title XX of the Social Security Act*

13

EPILEPSY

Epilepsy may be defined as "two or more seizures or convulsions that occur in the absence of an acute disease or injury to the brain and in a child who does not have an elevated temperature." Many people still think of epileptics as falling to the ground, their entire body stiffening, mouth frothing, and all four extremities jerking. This is only one type of epileptic seizure (grand mal). There are many other kinds of seizures. One involves only a brief staring spelling and loss of contact with the environment (petit mal); another involves movements of just a part of the body (focal seizures); and still another consists of unusual sensations or behaviors (psychomotor seizures). Table 13.1 summarizes the traditional and the new terminology for classifying seizures.

Some children may have only two seizures during their entire life, while others may have hundreds each day. Many children have seizures that go unnoticed by those around them. Only afterward, if ever, is it recognized that the child has been out of contact with what has been going on. Some children with epilepsy have no other neurological problems, but others suffer from retardation, cerebral palsy, learning disabilities, behavioral problems, or language disorders.

Since there are so many different types of epilepsy, it is not surprising that other terms have come to be used as synonyms for this disorder. These include "fits," "seizures," "convulsions" and even "fainting attacks." Since a fit, a seizure, or a convulsion is a single event, a child could have one such experience and not have epilepsy. If a child has had two or more seizures or convulsions, without a history of an acute disease or injury to the brain and without an elevated temperature, the terms seizures and convulsions mean epilepsy.

Table 13.1. Classification of epileptic seizures

Traditional terms	New international classification	Description
Focal seizures	Partial seizures with elementary symptomatology (seizures that begin focally)	These seizures may involve only one part of the brain and the symptoms and signs thus can be limited to one part of the body, i.e., repetitive twitching of the left corner of the mouth or the right thumb; numbness of the right hand; dilatation of a pupil or excessive sweating. Such seizures can be motor, sensory, autonomic, or a mixture of these (compound). They may spread locally to contiguous brain area, thus involving adjacent parts of the body in the seizure (a Jacksonian march), or they may spread to involve the opposite side of brain (secondary generalization). Unless there is secondary generalization consciousness is preserved.
Temporal lobe or psychomotor seizures	Partial seizures with complex symptomatology	Generally there is impairment of consciousness. This may be the only symptom. There may be associated confusion, inability to use or understand language, a memory lapse while a repetitive motor behavior continues to be performed normally but without purpose. It cannot be interrupted usually until the seizure is completed. Other motor acts may be stereotyped but without meaning (lip smacking, rubbing an arm, picking at one's clothes). These are automatisms. Fear, anxiety, and, rarely, rage can be the only seizure manifestation. Hallucinations (usually of smells or tastes, and less often of sounds) occur, as do unwarranted feelings of everything being strange and unfamiliar, or that new places or people are uncomfortably familiar (jamais and déjà vu). Thus, cognition, affect, sensation, and motor activity can be affected. These seizures begin in

Table 13.1. (*Continued*)

Traditional terms	New international classification	Description
		the temporal lobe but they too may generalize.
Hemiclonic seizures	Unilateral seizures	One side of the body is involved in clonic jerking or tonic stiffening. Consciousness may or may not be preserved.
	Generalized seizures (no focal onset)	
Petit mal seizures	Absence	Consciousness is interrupted with or without stereotyped movement. Posture is preserved.
Grand mal seizures	Tonic Clonic Tonic-clonic	The child loses consciousness and falls to the ground in most instances. There may be a period of rigidity of all muscles, repetitive rhythmic jerking of muscles or rigidity followed by jerking before consciousness is regained.
Infantile spasms	Bilateral massive epileptic myoclonus	The infant or child has a series of discrete jerks in which head and legs flex or extend and arms adduct or extend.
Minor motor seizures	Atonic Akinetic Myoclonic	Sudden loss of tone of neck or limb so that it drops. Sudden loss of truncal tone so that the child falls to the floor. Sudden shock-like jerks of a muscle or group of muscles sufficient to move part or all of the trunk, face, or extremity. It can be limited to one group or shift from one part of the body to another.

THE FAMILY PHYSICIAN

Many children who have had one or more seizures can be cared for by a family physician. Often seizures can be well controlled by a single drug given for a limited period of time. The medication most often used initially to treat children who have a single generalized convulsion or repeated febrile convulsions is phenobarbital. The seizures of many patients with grand mal, petit mal, or psychomotor epilepsy can be entirely suppressed or significantly reduced with one or at most two drugs. Most physicians are familiar with the use of phenytoin (Dilantin). At present, phenytoin and phenobarbital are probably the two drugs most commonly used for the treatment of either grand mal or psychomotor seizures. Most physicians in family practice are also experienced in using the succinimides (Zarontin and Celontin) for the treatment of petit mal.

CASE 13.1

RN, a five-year-old boy, had three grand mal seizures over a period of two months before being placed on phenobarbital. The seizures were not associated with fever. The family physician found no abnormalities on physical examination, nor was there anything in the child's history to explain the attacks. A brain wave (electroencephalogram or EEG) showed some slowing but no seizure activity at rest. The parents felt that epilepsy was a serious disorder that should be cared for by a specialist. The physician suggested that the family see a child neurologist in a nearby city, some one hundred miles away. A visit to the neurologist yielded no further information. He too found no abnormalities on physical examination, and another EEG showed general slowing but no paroxysmal activity at rest. The child had had no further seizures since beginning treatment and had no behavioral problems. The specialist suggested that the family physician could handle this problem but that he would see the child on subsequent occasions if the family doctor so desired.

Comment:

There is no reason why the family physician could not care for this problem assuming that the child continued to do well. Epilepsy can be a serious disorder or the symptom of an associated serious illness such as a brain tumor, but in RN's case there was no evidence to indicate that this was so. At least 60 percent of children who are otherwise entirely normal can be rendered free of generalized (grand mal) seizures with medication, and many of these children remain seizure-free after the medication is discontinued. Presumably, this family practitioner sought the advice of a medical specialist primarily to relieve the anxieties of the family. After having done this, it was appropriate that the family physician continue to follow the patient. Such care is more convenient and less expensive for the family and often is in the best interest of the child.

Children whose seizures are easily controlled with medication often do not have further medical problems. If they do have problems, these are generally related to social, educational, and vocational matters. Relationships within the family may also be a source of difficulties. The family practitioner who cares for the entire family, and often for the friends of the patient, is far better able to sort out these problems than a specialist who practices some distance away and sees the family only briefly on a few occasions. If other symptoms arise in the future, the specialist can again be called upon for advice.

If a child has a seizure disorder, the physician who looks after him may choose to begin treatment without consulting a specialist in neurology. In that case, he should be able to perform a neurological examination that is sufficiently detailed to assure that there is no accompanying neurological disease. He should prescribe antiepileptic medications with which he is familiar. It is very important that he discuss the possible toxic effects of these medications with the family. One of the most common problems encountered in the treatment of children with one or more seizures involves the use of phenobarbital. Although this is a very safe drug, in a number of small children it produces behavioral disturbances, including irritability, aggressiveness, and diminished attention span. If parents are not told that this may be an effect of the drug, they may feel that their child has a much more serious disease that is now affecting his behavior as well as producing seizures. When drugs are prescribed that may decrease the production of blood cells (e.g., phenytoin, Tegretol, Zarontin, Tridione, Mesantoin) or disturb liver function (sodium valproate [Depakene]), even though these complications may be extremely rare, parents must be informed. The family and patient should also be told how to recognize a drug allergy. These are also rare, but they can be extremely serious if not treated immediately. Such allergies generally begin with a skin rash; if the offending drug is withdrawn immediately the problem may be limited to the rash, which will gradually disappear. If the drug is not discontinued the child may develop a fever, severe sore throat, and eventually go into shock and die. Any physician who undertakes to treat a child's seizures must be able to direct the rapid withdrawal of the medication under these circumstances. Rapid withdrawal of an anticonvulsant drug often leaves the child more susceptible to seizures unless another drug is substituted promptly. Initially, this second drug may have to be given in relatively large doses for one or two days.

The physician will probably want a child to have an electroencephalogram (EEG) after his first seizure or if his seizures recur frequently.

The electroencephalogram records the electrical activity going on in the brain. This activity is usually rhythmic and symmetrical over the left and right sides of the brain. Abnormal EEGs can be too slow over all parts of the head or in one area (asymmetric, lateralizing, or focal). EEGs that are slow in one area suggest that seizures may be caused by some structural damage to brain, such as a head injury, stroke, or tumor. Other abnormalities are those of rhythm. Abrupt changes in rhythm with very fast sharp waves are called paroxysmal and suggest that the child is more apt to have seizures than are individuals who do not have this type of activity. Particular types of seizures may have unique EEG patterns. All EEG testing is done in a special laboratory and is interpreted by experts in this field. Their report is sent to the physician, who will explain the meaning of the electroencephalogram to the family and relate it to their child's seizure activity.

CASE 13.2

RC, a four-year-old child, had two generalized seizures without fever before being seen by his pediatrician. The pregnancy and delivery had been uncomplicated, and RC's development was normal. The patient showed no abnormalities on general physical examination or neurological examination. The pediatrician ordered fasting blood sugar and electrolyte tests, since low sugar, sodium, or calcium can cause seizures. Results of both were normal. An EEG was mildly slow. There were no focal abnormalities in the tracing, nor was there any paroxysmal activity.

The family had many questions. They wanted to know the cause of their child's seizures, and whether they could be sure he would not have seizures if he were treated with anticonvulsants. On the other hand, could they be sure he would have seizures if he were not placed on medication? They asked whether his epileptic disorder would get worse. Finally, they were concerned about the significance of the slow EEG that did not show paroxysmal epileptic activity. Did a slow EEG mean that their child would develop slowly?

Comment:

These are questions often asked by parents of children with epilepsy, but they are difficult to answer. When a child's history and physical examination reveal no abnormalities, the cause of the seizures often is never determined. In general, it is best to be thankful for this, since if the cause can be determined it generally implies a more serious problem, such as a tumor or an injury to blood vessels. Physicians usually assume that epileptic attacks that begin without an obvious cause are due to old scars resulting from a brain injury that may even have predated the child's birth. However, no one can be certain of this.

There is no way to predict which child will have more seizures with or without treatment. Within certain limits (based on statistics), one can say that a child with generalized seizures but with no evidence of other neurological disease has the best chance of having his seizures well controlled by drugs. However, statistics derive from large populations; they do not help predict the course of the individual patient. It is possible that a child may remain seizure-free without anticonvulsants, but if he has had two or more generalized seizures his chances of remaining seizure-free will be much better if he is treated with drugs.

Questions about the EEG can be answered by any physician who has some knowledge of this test. All physicians should have the ability at least to interpret to parents what an EEG report means with regard to a child's intelligence or the possibility of his developing other neurological problems in the future. About 15 percent of normal children have slightly slow EEG records; about 80 percent of people with epilepsy have the same EEG abnormality between seizures. A slow EEG is the most common abnormality recorded, but it is of little use in predicting the recurrence of seizures and it has nothing to do with the child's intelligence.

The doctor should be prepared to discuss the restrictions that may have to be placed upon a child's activity. We believe such restrictions should be minimal—in most instances, limited to certain activities that include climbing to heights, swimming, or taking baths in the absence of an adult. Other doctors may have different ideas about these matters and should discuss them with both the child and his parents.

Families of older children and adolescents may need guidance about educational, social, and vocational activities. If the patient is a teenager, the physician should certainly be prepared to discuss driving restrictions imposed on individuals with epilepsy in many states. He should also be prepared to work with a teenager in planning his future vocation. Many teenagers need to be reassured that epilepsy will not restrict their future sexual activities. Some are concerned about the genetics of epilepsy. If the cause of an epileptic seizure disorder is not clearly established, there is a slightly higher chance that the individual's children will also have seizures. (About 2 percent of children who have one parent with epilepsy will have seizures. The incidence in the general population is about 0.6 to 1.0 percent.)

The physician should have some plan for withdrawing a child from medication in the future. This should be discussed with the family and the child if he is old enough and has the capacity to understand these concepts. These responsibilities are not outside the competence of a family physician involved in the continuing care of patients with seizures.

As a child grows older, educational, vocational, or behavioral problems may require consultation with other professionals such as psychologists, teachers trained in special education, social workers, or psychiatrists. Any physician who undertakes the continuing care of a child should have established contact with practitioners in each of these disciplines.

There are several reasons why a family physician might want to refer an epileptic child to a specialist in pediatric neurology: for a second opinion to establish or confirm the diagnosis; for a thorough examination to rule out an underlying lesion such as a brain tumor; for advice about medication or diet if seizures prove difficult to control; for advice about stopping medication; or at the request of the parents if they wish a second opinion.

THE MEDICAL SPECIALIST

If a child is thought to have had a seizure or is suspected of having recurrent seizures, the specialist most likely to be consulted is the child neurologist. Complex cases may later develop problems that need to be treated by a neurosurgeon or a psychiatrist.

A family should expect to have the following information when the pediatric neurologist's evaluation is completed:

1. They should be told whether the child has had multiple seizures without fever (epilepsy). The parents should be informed if the child has partial (focal) seizures and if so whether it is possible to identify which part of the brain the disorder involves. They should be told if the child has been experiencing more than one type of seizure. In most cases a specialist should be able to distinguish between epilepsy and other recurrent attacks, such as fainting episodes, migraine attacks, low blood sugar, or anxiety attacks. A specialist should be able to decide if unusual behaviors such as rage attacks or sleepwalking are seizures.

The diagnosis of a seizure disorder is made primarily by obtaining a description of the spell from the child and his family or from others who have seen the attack. Often, the evaluation of these descriptions requires an experienced neurologist, who has had many past opportunities to listen to and evaluate such accounts, to determine whether the child has epilepsy.

If the description of the attack does not lead to a specific diagnosis, the most helpful laboratory test in determining whether a child has a seizure disorder is the electroencephalogram. In most instances, the child will not have a seizure while the EEG is being recorded. The child neurologist will know what special procedures are apt to bring out seizure activity if none is present in the routine record. Overbreathing, flickering light (photic stimulation), or sleep deprivation may help produce a meaningful record. A complete EEG can be done without hospitalizing the child but may take as long as two hours.

An experienced pediatric neurologist should be able to explain what the EEG means as it relates to a child's clinical history. It is important that the parents understand that the EEG has no meaning apart from their child's clinical problem. Thus, if a four-year-old child has suddenly fallen to the ground three times in the last two weeks, stiffened, turned blue and then begun to have jerking movements for the next several minutes, after which he falls into a deep sleep, he has epilepsy even if his EEG is normal.

2. Many times a specialist will not be able to tell parents what is causing the child's seizure disorder unless the seizures are associated with a clearly defined brain injury or disease, such as meningitis. Families often ask if a child who has epilepsy has permanent brain damage. Epilepsy generally means that a group of cells in the brain are not functioning normally. Thus all epilepsy can be said to be a symptom of brain damage. However, only a minute portion of the brain may be damaged and apart from occasional seizures there may be no other signs that the child is not entirely normal neurologically.

Even if a physician cannot identify the reason for the child's seizure disorder, he should be able to reassure parents that it is very unlikely that their child is suffering from some progressive problem such as a brain tumor or degenerative disease that will result in gradual loss of neurological functions. Frequently it is possible for a pediatric neurologist to give such assurances without admitting the child to the hospital. If the child is perfectly well apart from having generalized seizures and he has a completely normal physical examination, an EEG and a few simple blood tests done in the office may be all that is necessary to rule out an underlying disorder. On the other hand, if the child has seizures that appear to begin in one part of the brain (focal seizures), if his physical examination suggests significant brain damage, or if the brain wave test is abnormal in one area of the brain, the physician may want to do further tests. These may include sampling the fluid that surrounds the brain (a lumbar puncture or spinal tap) and special x-ray studies that can help determine whether there is an infection or tumor of the brain. Biochemical tests may be ordered to see if the child metabolizes sugar, fats, and amino acids (the constituents of proteins) appropriately, since seizures can occur in children whose body chemistry is abnormal.

Depending on the number and complexity of the tests needed, the physician may wish to hospitalize the child for several days. No matter how many tests are performed, however, it is impossible to give parents an absolute guarantee that their child does not have a very small tumor or abnormality of blood vessels that cannot be detected by the available techniques. Time is the only guarantee that a child who has recently developed epilepsy does not have a progressive disease or tumor. If over a period of several years he remains normal except for an occasional seizure, the chances are very slim that the seizures are symptoms of another disorder.

3. Pediatric neurologists should be familiar with all drugs used to treat seizure disorders. Commonly used drugs such as phenobarbital and Dilantin (phenytoin) are known to all physicians. In recent years, however, several new drugs, such as valproic acid, clonazepam, and carbamazepine, that are effective in controlling different types of seizure disorders have been approved by the Food and Drug Administration. Family practitioners may be less familiar with the use of these new drugs. If a child's seizures are difficult to control with commonly used medications or if he needs multiple drugs, the family physician may want a pediatric neurologist to prescribe the medications.

While it is the doctor's responsibility to prescribe medication, it is the parents' responsibility to decide whether a child should take antiepileptic drugs and which drugs they want him to take. A physician, be he in family practice or in a specialty such as child neurology, gives advice. He gives the parents his opinion about the need for treatment and suggests the therapy that he thinks best for the child. The parents make the final decision.

A physician must explain to the child's parents the possible toxic effects of any drug before prescribing it. He should tell the parents whether the toxic effects are common or rare and whether repeated laboratory tests will be needed to detect toxicity before the child has symptoms. Specialists are more familiar with the adjustment of drug levels for better control of seizures and to help avoid side effects. Drug levels need to be interpreted in relation to continued seizure suppression or frequency, side effects such as lethargy, and the potential hazards of the drug.

A pediatric neurologist will often see a child to help decide whether antiepileptic drugs should be discontinued. It is unusual to stop antiepileptic medications until a child has been free of attacks for at least two years. Some doctors require four to six years without seizures. The lack of agreement between doctors about this issue stems from the lack of dependable information about the frequency of seizure recurrence after drugs have been withdrawn relative to the time that a child has remained seizure-free while on medicine. In general, the longer a child remains free of seizures, the less likely they are to recur. Parents should always be warned not to remove a child from medication abruptly, i.e., not to run out of the prescribed medicine or to stop the medication without consulting the physician. Abrupt withdrawal of drugs makes the child more susceptible to seizures. If the drugs are withdrawn slowly over a period of weeks or months, the child is less likely to go through a period of increased susceptibility. If the doctor feels that a child should be removed from antiepileptic drugs, he should discuss the possibility of the recurrence of seizures with the parents and the child. There is evidence that children who have no other neurological abnormalities are less likely to develop seizures when medication is withdrawn than children who do have neurological deficits.

Older children who have their driver's licenses should be warned that in most states if they do have another seizure they will not be allowed to drive a car for a period of time. Teenagers who are working should also be alerted to the possibility that a recurrence of seizures could interfere with their employment. Children who have been seizure-free on medication face at least a 30 percent risk that seizures will recur when medications are withdrawn. On the other hand, children should not be treated indefinitely with potentially toxic antiepileptic medications because of the possible hazard of having another seizure. Unfortunately, there is no way of identifying which child will have a seizure when removed from drugs and which child will remain seizure-free. Most recurrences occur in the first two years after withdrawal of medicine, but they may come at any time in the future.

4. About 15 percent of children with seizures do not respond to any combination of drugs available. In another 25 percent, the seizures, although decreased, are never fully suppressed. Some of these children, particularly those under five years old who have sudden jerks or falling attacks, may respond to special high-fat diets. Special

diets may also have toxic effects and should therefore only be prescribed and monitored by a physician who is familiar with their use. Children who have focal seizures that cannot be controlled by drugs may respond to surgical removal of the focus if that area of the brain appears to have no other important sensory, motor or speech functions. Children are generally not considered for surgery unless their seizures are frequent enough to seriously interfere with their lives. The process of identifying which children might benefit from this type of surgery is highly technical and involves complicated tests to prove that there is such a focus of abnormal electrical activity in the brain. Surgical treatment is best undertaken at a medical center where there are neurosurgeons and neurologists on the staff who are experienced in the use of these techniques.

5. Children with epilepsy often have problems performing at the expected level in school. The reasons for this are not clear. They may also have difficulty gaining the acceptance of their peers. Later in life, the children may have vocational problems if the unpredictability of their seizures places restrictions on the kind of job they can perform. It may be dangerous for them to work with machinery, or they may not be able to obtain a driver's license to get to and from work. Further, the fact that they are in a sense "different" may lead to psychological problems. A neurologist caring for an epileptic child tries to anticipate these psychological difficulties by limiting the child as little as possible. No doctor can guarantee a family that their child will not have a seizure while riding his bike, roller-skating, or skiing. Nor can the doctor assure the family that the child is not having seizures in his sleep that they are unaware of. He can, however, assure a family that if they hover over their child constantly because he has seizures, or if they restrict his activities severely, he will grow up to be psychologically disabled. As noted earlier, physicians suggest the family place minimal restrictions on physical activities of the child. It is often suggested that contact sports, where there is a risk of head trauma, be avoided. If these sports are essential to the child's image of himself, however, the restriction may have to be relaxed. Some individuals are extremely susceptible to seizures when they breathe heavily and rapidly. If this is the case, the child must be introduced to strenuous exercise gradually in order to build up his tolerance. Parents should expect a specialist to discuss these problems with them and, if he does not, they should question him in detail about the activities he feels their child can engage in.

School problems due to frequent seizures may be relieved by manipulating the child's medication. Such problems often result from drowsiness caused by drugs and require that some compromise be made between the child's level of alertness and the number of seizures he is having. *The major goal of the physician who treats a child with epilepsy is usually not total control of the seizures.* Rather, it is to achieve a level of control that will allow the child to do as well as he possibly can in school and to interact with his peers normally. This means that the number of seizures may have to be balanced against the effects of medication at each stage of the child's development. It is the hope of every specialist caring for children who have epilepsy to eliminate the seizures completely while using the least amount of medication. However, it would seem preferable for a child to have an occasional focal seizure or even one or two brief staring spells a day and be alert and vigorous rather than to be seizure-free but drugged and unable to learn or participate in social activities. Many families of children with epilepsy are reluctant to accept this goal and don't understand why their child's seizures cannot be totally controlled. Often they fear the seizures far more than the child does. However, the parents of epileptic children have only a limited number of treatment choices, all of which involve risks. Seizures are a dramatic risk, but ultimately they are likely to be much less harmful than rearing a child whose intellectual social, and emotional growth has been stunted by overmedication or over-protection. Families who demand complete suppression of their child's seizures before allowing him to live a normal life may need professional counseling to help them adjust to their child's illness.

Vocational problems can be serious for adolescents with epilepsy, particularly if they are not interested in a college education or an office job. Preparation for future independence will require the skills of a number of professionals apart from the family physician and child neurologist. These may include psychologists, social workers, and counselors experienced in educational and vocational placement. Parents should expect that a specialist caring for their child on a continuing basis will obtain sufficient information about the child's abilities so that he can discuss these problems knowledgeably and refer their child to appropriate agencies.

All parents at times do not like the activities their teenage children engage in. Parents of epileptic children often use the child's disorder as an excuse to limit his activities and they try to enlist the doctor in this effort. This places the physician in an awkward position. As the child with epilepsy becomes an adolescent and finally an adult, he should feel that he can trust his physician and be comfortable speaking to him privately. His physician should not be considered an extension of his family. Nor should having epilepsy be used as an excuse to either forbid or fail to discipline a child for undesirable social activities. Parents must deal with these activities as they would if the child did not have epilepsy.

As children with epilepsy become teenagers, they often want to talk with a specialist or their family physician without their family being present. They may have concerns that are highly personal and their wish to speak to the doctor alone should be respected. These youngsters are often seeking information about the risks they face if they join their peers in experiences with drugs, alcohol, or sex. Alcohol should be restricted to an occasional weak drink, since there is little doubt that it can promote seizure activity in many individuals with epilepsy. Physicians generally caution very limited use of marijuana if any at all, though it is probably not as likely to increase seizure activity as alcohol. There is no medical reason why a teenager with epilepsy cannot engage in the same sexual activities as other members of his peer group. Nevertheless, it should be pointed out to children with epilepsy that physical exhaustion or prolonged lack of sleep do increase the chances of having a seizure. A girl who has begun to have intercourse should be warned that a number of antiepileptic medications may have an effect on the fetus should she become pregnant. If she is planning a pregnancy or thinks she is pregnant, she should inform her doctor immediately so that her drugs can be adjusted. Barbiturates, such as phenobarbital and Mysoline, are probably the least toxic to the fetus, although we still know little about the risks connected with the use of the anticonvulsants.

As they reach adulthood, individuals also want to know about the chances of transmitting epilepsy to their offspring. If the seizures are due to a known cause, such as a head injury, the chances that any of their children will have epilepsy is probably not greater than that of the general population (0.6 to 1.0 per hundred). If they have idiopathic epilepsy (i.e., there is no known cause for the disorder), the chance that their children will have epilepsy is probably two to three times that of the general population, but there is still approximately a fifty to one chance that the offspring will not be epileptic. If two epileptics marry, the risk of epilepsy in any child they have rises to about 10 percent. If they have a seizure disorder that is associated with an inheritable disease, such as tuberous sclerosis or neurofibromatosis, the chance that their offspring will have seizures is significantly greater.

THE PSYCHOLOGIST

The psychologist offers a wide range of services that can be helpful in the care and management of an epileptic child. These include evaluation of intellectual function, academic progress, and social and emotional adaptation.

However, special tests or special techniques to evaluate the overall performance level of a child with epilepsy are generally unnecessary (see Chapter 5). The psychologist has a unique opportunity in the course of an evaluation to observe the child's behavior and may be able to provide information about lapses of consciousness or other types of seizures that interrupt the child's attention and, as a result, interfere with learning.

Epileptic children and their families may profit from advice about behavior modification or family counseling, as discussed in Chapter 6.

THE EDUCATOR

Like any other group of students, children with epilepsy vary in regard to intellectual abilities. Most attend regular classrooms with their peers. If an epileptic child is placed in a special-education setting, the placement should be based on his abilities, not his disorder. Unfortunately, many teachers and other school officials are grossly misinformed about epilepsy and may resist having a child with a known seizure disorder in their classroom. Ignorance about the disease and what to do if a child has a seizure in the classroom, apprehension about the effect on the child's classmates if this should occur, and pressures from other parents who complain (upon learning from their child that a classmate had a "fit") that they do not want their children to witness a seizure, may lead teachers or principals to push for "special placement" of the child.

Some parents do not want anyone to know that their child has epilepsy and deliberately avoid informing school personnel. Such behavior not only reinforces the notion that there is something shameful about epilepsy; it may also provoke school officials to move for a special school placement if it is discovered that the child has seizures or if someone reports that he is taking "drugs" and the resulting investigation reveals that the child has epilepsy.

There are other reasons why teachers need to know about the child's disorder: (1) so that they will know what to do if the child has a seizure at school; (2) so that medications, if necessary, can be supervised by the school nurse or the teacher; (3) so that if a child is having a learning problem, teachers can try to observe whether this is due to sluggishness from overmedication or to small "spells"; and (4) so that they will know about any restrictions on the child's activities at school. Decision about limits on certain activities should be made by the child's physician and communicated by him to the school.

Thus, it is particularly important for parents of a child with seizures to establish good communication with their child's classroom teacher. The child's teacher will play a primary role in creating a supportive environment that will help him to feel secure in school. Moreover, the teacher's attitude influences the attitude of the child's classmates toward him. Table 13.2 is a sample information form that

---

Table 13.2. School information form for the student with epilepsy

---

Student's name _____ Birthdate _____
Sex _____
Parents_____ Phone_____
Address_____

Physician_____ Phone_____
Address _____

Type or description of seizures_____
Is the child currently receiving treatment?   _____ Yes _____ No
Type of medication taken _____
Time medication is taken_____
Is medication to be taken at school?   _____ Yes _____ No
If "Yes," has parental authorization been obtained?   _____ Yes _____ No
Who will be dispensing the medication at school? _____
Possible side effects _____

Likelihood and frequency of seizures during school hours_____

Any special instructions from physician_____

Parents' comments _____

provides professionals at school with necessary information about the child with epilepsy. Teachers may have the opportunity to observe seizures while the child is in class performing structured acts in a continuous manner. They can in turn provide important information to the doctor or the family by describing the attacks (Table 13.3).

Parents of a child with a seizure disorder should discuss the following points with the child's teachers:

1. Briefly and clearly describe the type of convulsion the child experiences.

2. Explain that with most seizures, the child is *not* experiencing any pain and that once a seizure has started it cannot be stopped.

3. Describe what is done at home when the child has a seizure. For example, lay the child on his side, loosen his collar, *do not* try to force the child's mouth open, *do not* place anything in the child's

---

Table 13.3. Anecdotal record for the student with epilepsy

Child's behavior before the seizure (precipitating events, if any)? _____
_____

Child's behavior during the seizure (sides and parts of the body most affected during the seizure)? _____
_____

Child's behavior after the seizure? _____
_____

Were you able to note the time of seizure?_____
Reaction of other students_____
_____

Any injuries sustained as a result of the seizure?_____
_____

Does the child exhibit any unusual behavior that could be a resulting side effect of anticonvulsant medication (drowsiness, inattention, irritability, etc.)?_____
_____

Comments:_____
_____
_____

mouth. For the child's emotional as well as physical security, it is important that the teacher's approach is consistent with the approach used at home.

4. Emphasize the importance of remaining *calm*. Explain that there is very little the teacher can do for the child who is having a seizure.

5. Finally, parents should point out that a calm and matter-of-fact handling of a seizure by the teacher will influence the other students' attitudes toward the epileptic child. An article published by the Comprehensive Epilepsy Program for the State of Minnesota suggests: "Other children may react with surprise, fear, or revulsion to a seizure. School personnel can be very instrumental in preventing any lasting negative reaction on the part of other students. If the children are old enough to understand, the nature of epilepsy should be explained. If not, a simple reassurance that it does not hurt and will soon pass will suffice. To the child with the seizure, it is vitally important that continued acceptance be assured through the school person's calm, open behavior."

The school nurse should be informed of the child's condition and what, if anything, she needs to do if the child has a seizure. If the child has a single, brief seizure, it is best for him to remain in school. Parents should inform both the nurse and the teacher that they prefer that the child remain in school after a seizure. If he is drowsy, as is often the case, he can rest briefly in the aid station and then return to class. A change of clothes should be left at school for the child's use if he has a tendency to wet or soil himself during a seizure. Parents should also brief other people at school, such as the bus driver and lunch room supervisor. The Epilepsy Foundation of America has put together an excellent packet of materials entitled "Epilepsy School Alert." These free materials designed for classroom teachers, school nurses, and administrators explain many aspects of epilepsy and are available upon request.

REFERENCES

Aird, R. B., and D. M. Woodbury. 1974. The management of epilepsy. C.C. Thomas, Springfield, Ill., p. 448.

Epilepsy and insurance. 1975. Epilepsy Foundation of America, Washington, D.C., p. 55.

Freeman, S. W. 1979. The epileptic in home, school and society. C.C. Thomas, Springfield, Ill., p. 285.

Lagos, J. C. 1974. Seizures, epilepsy and your child. Harper & Row, New York, p. 273.

Livingston, S. 1972. Comprehensive management of epilepsy in infancy, childhood and adolescence. C.C. Thomas, Springfield, Ill., p. 657.

Magnus, O., and D. Lorentz, eds. 1974. The epilepsies, handbook of clinical neurology, vol. 15. P. J. Vinken and G. W. Briegu, eds. North Holland Publishing Co., Amsterdam, p. 860.

Meinardi, H., and A. J. Rowan. 1978. Advances in Epileptology, 1977. Sivets and Zeitlinger, Amsterdam, p. 468.

Penry, J. K., ed. 1976. Epilepsy: Eighth international symposium. Raven Press, New York, p. 414.

Pippenger, C. E., J. K. Penry, and H. Kutt, eds. 1977. Antiepileptic drugs: Quantitative analysis and interpretation. Raven Press, New York, p. 400.

Report of the panel on convulsive and neuromuscular disorders to the National Advisory NINCDS Council. 1979. DHEW Publication No. (NIH) 79-1913, p. 124.

Sands, H., and F. C. Minters. 1977. The epilepsy fact book. F.A. Davis Co., Philadelphia, p. 116.

Salomon, G. E., and F. Plum. 1976. Clinical management of seizures: Guide for the physician. W.B. Saunders Co., Philadelphia, p. 152.

Svoboda, W. B. 1979. Learning about epilepsy. University Park Press, Baltimore, p. 240.

U.S. Commission for the control of epilepsy and its consequences: Plan for nationwide action on epilepsy. 1977. Department of Health, Education and Welfare, National Institutes of Health, Bethesda, Md. DHEW publication nos. (NIH) 78-276, 78-277, 78-278, 78-279, 78-311, 78-312.

Ward, F., and B. D. Bower. 1978. A study of certain social aspects of epilepsy in childhood. Develop. Med. Child Neurol., Supplement 39 to Vol. 20:1–63.

Wright, G. N., ed. 1975. Epilepsy rehabilitation. Little, Brown, Boston, p. 275.

14

CEREBRAL PALSY

Cerebral palsy refers to problems with control of movement resulting from an injury to the brain before or around the time of birth. This group of symptoms may take many forms. The child may have difficulty moving because he is weak or floppy (atonic or hypotonic, meaning decreased muscle tone at the joints) or too stiff (hypertonic or spastic, meaning increased tone at the joints). The child may also have writhing or twisting movements (choreoathetosis) which make him incapable of controlling his trunk or limbs appropriately. Or he may develop a twisted posture of limbs or body at rest or when he moves (dystonia). In a few children, cerebral palsy is manifested by an inability to control balance (ataxia). Some cases of cerebral palsy appear to get worse with age although the underlying lesion in the brain does not increase in size. This is because the child's symptoms and signs change as the uninjured parts of the brain develop. Normal development of some parts of the brain is required to bring out the abnormality of movement. About one-third of children with cerebral palsy have only motor problems. The remainder suffer from various other disorders in addition, such as mental retardation, seizure disorders, learning disabilities, or behavioral problems. The term cerebral palsy refers only to the motor disability. It is necessary to follow the development of a child with cerebral palsy to determine whether any of these associated problems exist. If they are present, their implications must be made clear to the family in order to plan for the child's future.

THE FAMILY PHYSICIAN

When a physician tells parents that their child has cerebral palsy, their first questions usually are: What caused it?; How bad is it?; and What will the future hold for him? These are perfectly reasonable questions that must be answered eventually. However, the most reasonable question of all is: "What is cerebral palsy?" "What exactly is wrong with my child?" Cerebral palsy, as we have noted, is defined as a nonprogressive disability in motor function due to an unchanging brain lesion present from intrauterine life or acquired at or near the time of birth. It is important for the family to know which aspects of their child's motor function are not developing appropriately. Is the difficulty confined to both legs, or does it involve all four limbs? Is it confined to only one side of the body? Is there a partial or total paralysis of movement? Is the child just stiff, or do involuntary movements interfere with those he wishes to make? Are his limb movements normal, but just unsteady when he tries to walk? "Cerebral palsy" is a broad diagnosis, and it includes many specific abnormalities of movement that must be treated differently. Parents should expect that their family physician will be able to tell them the exact abnormalities that are present or refer their child to someone who can.

However, a family should not expect their physician to predict the ultimate outcome of a particular abnormality of movement at the time of the first examination, especially if the infant is very young. Many abnormalities of movement that appear relatively severe shortly after birth, particularly spastic diplegia and hemiplegia, may improve dramatically in some children yet persist in others. It is often difficult to predict which way a child will develop. Other abnormalities, such as unintentional twisting and turning movements that interrupt volitional efforts, may not be apparent until the child grows older and his nervous system matures (see Chapter 1). A physician may suspect their presence on careful examination but not be able to determine their severity for several years.

Since cerebral palsy really refers to a disorder of movement, it is important to know whether there are other symptoms of brain damage. Most important among these are mental retardation or recurrent seizures. As we indicated earlier, it is very difficult to evaluate retardation early in life if the child has abnormalities of motor function. A family may have to endure a long period of watchful waiting before a diagnosis of retardation can be made or rejected with certainty. On the other hand, there is usually no problem in making the diagnosis of epilepsy. However, if no evidence of epilepsy exists at an early age, it is not possible to assure a family that it will not occur later in life. Children with cerebral palsy are more likely to develop epilepsy than children without a motor deficit.

Since a child who has cerebral palsy associated with seizures or mental retardation has a much worse social, educational, and vocational prognosis than a child who has cerebral palsy alone, parents should not be satisfied with the term "cerebral palsy" as a diagnosis. They should insist that their doctor use clearly defined, specific terms when he describes what is wrong with their child, e.g., spastic diplegia. If parents do not understand the meaning of any of the terms the physician uses, they should ask him to explain them more clearly. A physician who undertakes the continuing care of a patient with cerebral palsy should be willing to explain these matters.

The recurring question "What caused our child's brain damage?" remains difficult to answer. As with so many neurological problems, the cause is often not clear. If there were serious problems during pregnancy or delivery or if the child had yellow jaundice shortly after birth, physicians frequently calm the family by assigning responsibility for the cerebral palsy to that particular event. Such associations, while sometimes valid, often are not.

Making predictions—satisfying parents' desire to know what their child's future might be—depends on being able to determine the severity of the motor deficit and the presence of associated problems. Since this may take several years, a physician is apt to give a very guarded prognosis early in infancy. He may wish to consult a neurologist to get more detailed information about the physical handicaps that are present in infancy. Later in the child's development, he may wish to have a psychologist evaluate the child to obtain a better appreciation of his mental abilities. At that point, the physician will be able to give the parents a more precise prognosis.

The treatment of cerebral palsy, as with so many neurological problems, involves the collaboration of the physician with other professionals. It is important that the family physician be in contact with physical and occupational therapists who can provide appropriate services for the child. It is also necessary that he be aware of potentially crippling orthopedic problems that may develop in cerebral palsied children. The family physician should see the child at frequent intervals to check for evidence of curvatures of the spine, tightening of the joints, or shortening of one leg relative to the other. He should refer the child to an orthopedic surgeon if he feels the signs indicate it.

If the child is mentally retarded or suffers from epilepsy, those conditions will obviously require special planning or treatment (see

Chapters 13 and 16). A physician should be prepared to counsel parents about the limits of their child's physical abilities as the child develops. He should help parents cope with any social and behavioral problems that may arise. Children who are not free to participate in the activities enjoyed by other children do not have the opportunity to learn appropriate social skills. Advising parents about the type of training that their child needs in order to develop appropriate social behavior is a very important aspect of the physician's role.

Many families ask if cerebral palsy is inherited. By definition it is not.

CASE 14.1

AP was born prematurely (twenty-eight weeks of gestation). He weighed only 1200 gm (2 lbs 10 oz). Initially, the child was placed in an incubator and given extra oxygen. He had no problems during the first two weeks after birth except for occasional attacks when he stopped breathing. Each time he was quickly revived without any other incident. He was allowed to leave the hospital when he weighed 2200 gm (4 lbs 12 oz), eight weeks after birth. At the time of discharge the baby was observed to move his arms a great deal when agitated, though his legs showed relatively little movement and remained in a flexed position, much like that of a frog (frog-legged). Muscle tone in his legs was decreased although reflexes were quite active. Later, upon examination, his family physician also noted that the child's head was not growing at the expected rate. He was concerned and referred the child to a specialist, who confirmed the physical findings and noted that at five months after birth, the child would look at objects but he did not follow them fully across his visual fields. The pediatric neurologist indicated that the patient suffered from a spastic diplegia associated with premature birth and in the future would probably have some degree of psychomotor retardation as well, judging from the lag in visual development and the slow head growth. However, no definite prognosis was made about the child's visual or intellectual development.

As the child grew older, it was quite apparent that he had problems with social and cognitive development as well as with the use of his lower extremities. The child's mental and physical development were discussed frankly with the family. As the patient failed to progress, the parents were gradually prepared for the fact that the child might be moderately or severely retarded. The child was referred to a cerebral palsy day care program when he was ten months old. He initially received physical therapy. His development progressed slowly, but by eighteen months he was able to sit alone, reach for objects, bring them to his mouth, laugh, and babble. He did not transfer objects at that time. He did recognize his parents and could follow objects fully across his entire visual field. He appeared to hear sounds. His leg muscles were becoming tighter, and his parents were taught how to passively stretch the muscles at home. In addition, he was started on an occupational therapy program and the parents were also taught how to exercise him at home. When he was 2½ years old his family physician arranged that he enter a government-sponsored early childhood stimulation program, where he received more intensive therapy on a daily basis. At age 3½, he was referred to an orthopedist, who advised short-leg braces to aid with his walking. The child is now five years old. He is in an early education program for handicapped children. The physician considered this the best placement available at present, although testing by a psychologist at United Cerebral Palsy suggested that he was only in the trainably retarded range. Audiological tests revealed no evidence of a hearing loss and an eye examination revealed his vision to be normal. He has continued with occupational and physical therapy. He now walks, can feed himself with a spoon, has a fifteen-word vocabulary that includes several two-word phrases, and has recently been toilet-trained using a behavior modification program outlined for the family by the consulting psychologist.

Comment:

This child suffers from a form of cerebral palsy that is limited to poor use of the lower extremities (spastic diplegia) and is often not severe. Unfortunately, in this child's case the complications of prematurity with possible associated problems before birth resulted in mental retardation as well. The family physician was aware of all these complications and prepared the family for them. He made use of appropriate specialists in the community. The information and recommendations provided by the specialists involved were reviewed with the family at each stage in the child's development and, based on this advice, the family physician arranged placement in the treatment facility that offered the child the best advantage. The doctor made no attempt to break up the family unit and received no indication from the family on repeated visits that caring for this handicapped child was causing excessive emotional or other stress to the family. In this case, a complicated problem was very well handled by a family physician who was interested in this type of care.

THE MEDICAL SPECIALIST

Specialists involved in the diagnosis and treatment of cerebral palsy include child neurologists, orthopedists, geneticists, urologists, and occasionally neurosurgeons. Many pediatricians are interested in caring for these particular neurological handicaps and are attempting to form a new medical subspecialty, Developmental Pediatrics. However, this specialty is not yet recognized by a qualifying examination that is approved by the American Board of Pediatrics.

The child who is eventually diagnosed as having cerebral palsy is usually referred to a specialist because of delayed motor development. Older children may be referred because of poor balance or the development of involuntary writhing or twisting movements of the trunk or limbs. Often the initial referral is to a pediatric neurologist, although it may be to an orthopedist. Frequently, the diagnosis of cerebral palsy is made while the child is being evaluated for a different neurological handicap, such as mental retardation or epilepsy. During the child's physical examination the developmental motor disability is recognized by the physician.

If a child has cerebral palsy, a specialist should be able to help the family with several aspects of the problem.

1. He should be able to describe the exact physical disability. Cerebral palsy is a loose term encompassing many motor disabilities that may have profoundly different effects on the child's development. The element common to all these disabilities is the failure to develop normal use of the limbs or trunk because of brain damage *in utero* or around the time of birth. It is therefore a nonprogressive disorder of movement and/or posture. In infancy, children with cerebral palsy may appear to be loose-jointed and excessively floppy with poor muscle tone, although their reflexes remain active. As they grow older, muscle tone may increase (they become spastic), or they may develop abnormal, uncontrollable movements.

The most common form of cerebral palsy is one in which the child is spastic, i.e., there is increased tone to the limb and increased reflex activity associated with poor fine motor control and occasionally with weakness. Spastic cerebral palsy can involve two limbs on the same side of the body (hemiplegia), all four limbs (quadriplegia), or all four limbs with the lower extremities much more affected than the upper extremities (spastic diplegia). Rarely are the upper limbs involved to a greater degree than the lower limbs (double hemiplegia). The terms themselves are less important than the description of the defect in a particular child.

In other forms of cerebral palsy, the child may not be spastic but he will have a movement disorder. When writhing, sinuous movements predominate, the name choreoathetosis is given to the disorder. These movements may be more obvious in the hands and feet but often involve all four limbs. Twisting movements that lead to fixed, abnormal postures are generally described as the dystonic form of cerebral palsy. Less frequently, the disorder manifests itself by loss of balance. It is possible for all of these movement abnormalities to be combined in the same child. Cerebral palsy need not be confined to the trunk or limbs; it can involve the muscles of the face, mouth, and neck, making chewing and swallowing difficult and interfering with the development of speech. The child often drools and makes repeated grimaces.

While it is important for a child with cerebral palsy to be diagnosed early in life, it is equally important for the family to be aware that these disorders of motion and posture evolve with time, even though the brain damage that is responsible for the problem does not progress. This evolution occurs because the remainder of the brain is developing normally, affecting the way in which the damaged area expresses its function. Thus, a child who is very floppy in the first months after birth may begin to show signs of increased tone and spasticity with crossing (scissoring) of the legs between six and fourteen months of age. In some children, choreoathetosis and dystonia do not appear until the second year of life, and difficulties with balance may not be apparent until the child begins to attempt to walk. Certain types of cerebral palsy, particularly the mild form of spastic diplegia and mild hemiplegias apparent early in infancy, may actually improve during development to the point that they are no longer noticeable on examination and do not interfere with the child's function.

Repeated evaluations over a period of several years may be necessary to gauge the full extent of the disability and to plan a long-term program of care, although immediate goals can certainly be set at the time of the initial visit. Under ideal circumstances, a diagnosis should involve a functional classification in which the degree of limitation is identified, but this may not be possible until the child is somewhere between five and ten years of age. Even then severe limitations can sometimes be overcome with training and tremendous motivation if the child has adequate intelligence.

2. If a child has cerebral palsy, his parents will want the specialist to tell them its cause. He probably will not be able to do so, as he faces the same problems as the family practitioner in this regard (Chapter 12). Even with the use of the CT scan (computerized axial tomography), a special x-ray that gives an image of the brain and allows the neurologist to visualize defects in the development of parts of the brain much more easily than in the past, the cause of such defects still is often unknown.

3. Parents certainly want to know their child's prognosis. How will

he develop in the future? Will he be able to work? As we have indicated, an accurate prognosis may only be made after several years of observation and an evaluation of the child's response to treatment. Prognosis depends not only on the severity of the disturbance of movement and posture, but also on coexisting problems. The child with cerebral palsy who is also retarded or who has seizures that are difficult to control has a much less hopeful prognosis than one who has only a disorder of movement or posture. Defective vision or hearing also may impede the progress of the cerebral palsied child. These must be looked for by any physician, because they sometimes can be partially corrected.

Past experience suggests that less than half of the children who are diagnosed as having cerebral palsy will become effectively employed adults, and that about one half of these will be employed in unskilled jobs. Approximately another ten to fifteen percent will be able to participate in a structured setting, such as a sheltered workshop. In general, children who have spastic paralysis of one side of their body or spastic diplegia tend to do the best with regard to their ultimate prognosis for living an independent life. These are rather depressing statistics, but there are individual exceptions. Children with choreoathetosis or spasticity involving all four extremities and associated with speech impediments have, if bright enough and well motivated, gone on to college and become highly productive members of the community. Judgments about the future of any child with cerebral palsy have to be made on an individual basis, taking into consideration psychological factors as well as the severity of the child's physical and mental handicap.

4. Parents will want to know what treatment facilities are available to their child. The treatment of cerebral palsy is a complex problem that requires the skill of many professionals. For example, highly trained psychologists may be needed to evaluate the true mental abilities of a child who has a severe motor defect. Standard psychometric tests may not be suitable for a cerebral palsied child who cannot control his muscles well enough to make the responses necessary to demonstrate his abilities.

It is important to involve physical, occupational, and speech therapists very early in the course of the disability. As soon as it is clear that an infant may be spastic or developing fixed postures, parents should be taught how to passively move his limbs to avoid contractures. Therapists can work with young infants to stimulate appropriate swallowing and sucking; later in infancy and childhood, if necessary, therapists can help teach the child how to bear weight, first in the sitting position and then while standing. Special devices may be required to assist him, which therapists and physicians are knowledgeable about. Later the child may need to be taught how to transfer in and out of a wheelchair and to get up and down stairs. Physical therapists are trained to teach these skills (see page 122). The use of the upper extremities must be stimulated and fine control of movement enhanced. Occupational therapists are skilled in these matters (see Chapter 8). The speech therapist can help the cerebral palsied child improve his articulation by training him to coordinate the use of the muscles of his throat, mouth, and lips.

The advice of an orthopedic surgeon may be needed in order to prevent deformities of the joints or to loosen muscles that are so tight that walking or independent movement of parts of the limb are impossible. Children with cerebral palsy also are prone to develop curvatures of the spine that may require orthopedic treatment.

Cerebral palsied children often require the help of educators to devise programs that will meet their special needs (see page 211). These needs depend on the child's motor defect and his mental ability. Some children may be able to learn to type, though they may never be able to write. Others may need to have the aid of a page turner in order to read, although their reading skills per se will not be defective. The immense effort required to overcome physical handicaps may result in mood disturbances that will require the attention of a psychiatrist. Some handicapped children with cerebral palsy may develop behavior problems because they have not been taught to conduct themselves properly. They are frequently overprotected or overindulged by their families and not required to learn the same rules of good conduct as other children their age. If this is the case, the help of a psychologist or social worker may be needed. Social service agencies will almost certainly be required to direct the parents of these children to centers where needed services for their child are provided and, if necessary, to secure financial assistance for the family during these long years of training. Ideally, the child with cerebral palsy should be referred by the physician responsible for his continuing care to other medical professionals at appropriate stages in the child's development or as specific problems become evident. A medical specialist who takes over the care of a child with cerebral palsy should have a working relationship with other professionals and understand the services these individuals can offer the child as he grows into adult life. From the very first visit, the specialist should begin to serve as a channel to other skilled professionals who will be required to treat the child.

The specialist should also advise parents about the limitations of certain programs. For example, there is no evidence that programs that improve motor control or deliberately increase sensory stimulation in one or more modalities have any effect on ultimate intelligence or on learning school skills later in life if the child is normally stimulated in the course of the day. Many fortunes have been made by convincing parents that training in gross and fine motor coordination or improving eye movements will lead to a more intelligent child. Most physicians deny this.

THE PSYCHOLOGIST

Cerebral palsied children who have moderate to severe motor difficulties present a more complicated and challenging assessment problem than children with most other types of handicapping conditions. While many of the traditional standardized tests can be used cautiously in mild cases, the examination of children with moderate to severe neuromuscular disabilities is difficult. In some cases, communication is a problem and tests that demand verbal responses are useless. Some children who are physically unable to speak are also unable to point. A highly skilled examiner who has a solid background in psychological assessment and experience in testing handicapped children and who in addition has had an opportunity to observe the child in question before the evaluation may be able to develop a fairly dependable system of communicating with the child (e.g., eye movements, facial expression, touching the right leg for yes or the left leg for no) that will permit the use of such tests as the Peabody Picture Vocabulary Test, the Ammons Full Range Picture Vocabulary Test, or the Van Alstyne Picture Vocabulary Tests. These tests consist of cards bearing pictured objects or activities about which the child is asked questions. If the child is unable to speak or point, a multiple choice approach can be used with these tests. The examiner asks the questions, points to each of the pictures in turn and the child indicates his or her choice by signaling "yes" or "no" in the agreed-upon code. Although these tests help bypass the communication problem, it must be remembered that tests such as these sample only a limited aspect of intelligence. The Ravens Progressive Matrices is a test that has been found useful to assess the ability of cerebral palsied children to "see" relationships and to reason logically at a more abstract level than the picture vocabulary tests. This test consists of sixty pictured designs, each of which has had a part removed. Each design is accompanied by six or eight alternative parts of designs that could fit into the empty space in the larger design. The child must choose the missing insert from the alternatives offered. Like the picture vocabulary test, the Ravens can be administered as a multiple choice test.

The California Mental Maturity Scale (CMMS) was developed for individual use with cerebral palsied children. The CMMS is a pictorial classification test made up of ninety-two items, each consisting of a set of three to five drawings on a large card. The child is asked to identify the one drawing on each card that does not belong with the others. He can indicate his choice by pointing or nodding. The cards and drawings are executed in different colors, and the objects pictured were selected on the basis of the range of experiences of most American children. Scores are expressed as age scores (i.e., 4.5 or 6.0 years). This permits the psychologist to compare a child's performance with that produced by children his age who were included in the standardization population. The CMMS can be administered either orally or by pantomime. However, its use is restricted to children in the three- to ten-year mental age range, which means it is not useful for children who are mentally older than age ten. The most common approach to the assessment of children with cerebral palsy has been to combine selected items from traditional standardized tests and to adapt them as necessary to fit the condition of the particular child. This approach makes the assumption that all such items taken together will give at least a general idea of the mental level at which the child functions. In the hands of a well trained, experienced psychologist, reasonably accurate qualitative and quantitative approximations of the intellectual potential of the child can be obtained. If a child has a moderate or severe motor deficit, the parents should be sure the psychologist has not only had extensive supervised training in testing handicapped children, but is someone who has had considerable experience in testing and psychological assessment. If this is not the case, the chance of an inaccurate evaluation increases enormously. An inadequate psychological assessment can lead to unrealistic expectations about a child's future performance.

THE EDUCATOR

This discussion will be limited to those physically handicapped children of average or above-average intelligence with cerebral palsy, neuromuscular disorders, or a birth defect.

Presently, states are not legally bound to provide free infant and preschool programs for children between three and five years of age, but some do and may even offer services to younger children. Preschool programs funded by state and federal grants are increasing, even in states that do not mandate educational services for children under the age of five. Parents should check with their state and local school administrators for information about the early education programs offered in their state. Private facilities or centers funded by contributions collected by national organizations of parents sometimes offer such programs. Occasionally, it is possible to find early stimulation programs for children who are under one year of age. These programs are usually set up as parent-infant classes and a parent is required to attend with the child. Stimulation activities are stressed, such as helping the child become more alert to his environment by introducing him to toys that improve sensory awareness. Physical exercises designed to strengthen weak muscles are an integral part of such programs.

Preschool programs for children from three to five are usually more widely available than infant programs. Professionals from many disciplines are involved in such programs, including occupational, recreational, and physical therapists, and speech teachers. These programs stress the attainment of various self-help skills (eating, dressing, and toilet training), but attention is also given to language and speech development. The preschool teacher focuses on the development of readiness or preacademic learning skills such as auditory and visual discrimination that contribute heavily to a child's later success in the classroom. Not to be overlooked is the fact that enrolling a handicapped child in a preschool program also affords the child opportunities to interact and socialize with peers.

Once the child is of school age and ready to begin kindergarten, it is the responsibility of the local school district to make sure that he receives a free and appropriate education that meets his needs. One common difficulty that arises when physically handicapped children attempt to join a regular classroom is how to overcome physical barriers. Public Law 94–142 states that if it has been determined that a child's needs could best be served by attendance in a regular classroom, the school district must make *every* possible attempt to accommodate the child. Some schools have been able to accommodate children in wheelchairs by building ramps or switching classes from a second-floor to a first-floor classroom. School districts can apply for federal funds to modify school buildings so that all areas used by students are made accessible to children in wheelchairs. If ramps cannot be installed, arrangements could be made for a parent, teacher, custodian, or other helper to be available before and after school to help the child enter and leave the school building.

If parents feel that a school district is being uncooperative about making the necessary architectural changes so that their child can attend school in the environment appropriate to his educational needs, the parents may initiate a hearing to bring the school district into compliance. (Instructions about how to initiate a hearing must be included in the notification of placement letter sent to the parents by the school district following the IEP meeting.) Parents are permitted to have a medical specialist in neuromuscular disabilities (e.g., neurologist, orthopedist) or a nonmedical expert (director of United Cerebral Palsy or Muscular Dystrophy Association) present at the hearing to advise the school district about what is needed specifically to accommodate the child.

Since we are talking about children of average intelligence or above, there obviously will not be a need for any "special" education other than that provided to all children in the same grade. However, it is not uncommon for physically handicapped children to need some sort of adaptive device to help them hold a book or pencil. In school, it is the responsibility of the local school district to provide this equipment. If special equipment not available through the school is needed at home to permit the child to do homework, parents should contact the local chapter of one of the national organizations of parents of handicapped children for help. Many of these organizations will lend such equipment to handicapped children.

During the child's school years, assessment procedures used to measure progress will of course have to be modified to take into consideration his physical limitations. For example, if a child's major problems involve his upper limbs, it would be unreasonable to administer timed tests and expect that he could work as rapidly as a neurologically "normal" child. In such circumstances, parents should discuss the matter with their child's teacher or counselor. Examination procedures should be modified so that the child is not penalized by his physical handicap.

When a physically handicapped but otherwise "normal" child nears high school age, he and his parents should talk with the school counselor about beginning to plan for the youngster's future, whether it is to be vocational training or advanced education. If the child's intellectual abilities are near average or above, it is very important that he be included in the planning sessions so that he can express his desires. Plans for the child's future are apt to fail unless the child feels that he has had an active role in making them. The school counselor should be able to supply the family with the necessary information about specific colleges or universities (e.g., if their physical facilities have been modified so that a handicapped student can participate in campus activities). Parents should not be surprised if the school counselor proves to be negative about the notion of college for a handicapped student. Unfortunately, such attitudes toward higher education for handicapped students are common. Therefore, parents would be well advised to do the initial information gathering themselves. They need only write to the Office of Admissions of any college explaining their situation and the information they request will be sent to them. It would also be wise, if possible, to visit several colleges well before entrance exam time to make personal contact with the Dean of Admissions and an admissions counselor. These persons could prove valuable if it becomes necessary to ask for modifications in exam procedures. The important points for parents to remember are:

1. Be realistic about their child's abilities.
2. Include their child in all planning sessions.
3. Investigate possible college placement early.
4. Have copies of all pertinent information about their child.
5. Introduce their child and themselves to the Dean of Admissions.
6. Don't be discouraged by either high school or college guidance counselors.

If vocational training is decided on, the school counselor is a source of specific information. In addition, parents should contact the State Division of Vocational Rehabilitation for information about specific programs and for assistance with enrollment procedures (see Chapter 12 for a discussion of the Vocational Training Act).

THE PHYSICAL AND OCCUPATIONAL THERAPIST

Programs for young children and infants should use the parent or caregiver as the primary teacher or therapist whenever possible. Neither the child nor the parent will receive optimal help if the parent sits in a waiting room while the child is being treated or if the child is dropped off at a day care program for the morning and picked up after treatment. Parents should be instructed in therapy activities they should use at home with the child for feeding, bathing, positioning for sleeping, dressing, play, and all other activities during the day. By using therapy activities at home, treatment is extended to include the child's whole day. Therapy procedures need to be carried out throughout the day because no infant or young child will benefit maximally from thirty minutes of therapy once a day or three times a week. One-half hour of therapy will not overcome hours of poor positioning and handling. The therapist is responsible for teaching and involving parents in the child's therapy. Parents should be told what the child does at the center and they should be given a home therapy program to follow with their child. If parents feel that the services provided by a therapist do not fill their needs, they should discuss their concerns with the program's supervisor.

For some children it is important to begin therapy early in the first year of life to prevent development of contractures and other joint deformities, since an infant's muscle tone is usually less than it will be when he is older, and abnormal movement patterns are not as well established in infancy. Much of the stiffness and disability caused by the deformities seen in older cerebral palsied children may be avoided if treatment is begun early. Parents would be wise to seek a second opinion if their child's doctor treats their concerns lightly or puts them off, saying, "He'll walk when he wants to" or "Let's wait until he is older so he can work with the therapist."

Therapy should follow a normal developmental sequence, first encouraging normal postures and then patterns of movement by using purposeful actions. Movements should be repeated using different toys or objects so that the child is not bored. Many kinds of rewards can be used to encourage a child to participate in an activity. Food should not be used as a reward because obesity will make movement more difficult. Playing with a special toy, verbal praise, cuddling, and other actions that indicate approval to the child, or which he finds to be positive, can also be used to maintain his participation in an activity. The ultimate goal of any treatment program is to develop as much independence in motor abilities as possible.

Programs for young infants and the more severely affected child focus on developing head and trunk control such as propping on elbows or hands while lying on the stomach, and achieving and maintaining the sitting position or motor abilities such as reaching for toys and bringing the hands together. Therapy for the child with mild hemiplegia might concentrate primarily on shifting weight to the involved extremities, exercises that encourage the use of the involved side, establishing symmetrical postures, and activities that require bilateral coordination.

This is not a program that takes an hour in the morning and then can be forgotten the rest of the day. It should be incorporated into normal play activities, feeding, bathing, and other handling of the infant or child throughout the day. The responsibility for care of the handicapped child frequently falls on the mother and in some cases the maternal grandmother. We strongly suggest involvement of the father in treating the developmentally disabled child, not only to support the mother but also to establish a better father-child relationship. This provides the mother some relief and fosters more positive family relationships. No one profits from one person's martyrdom. If the child is taken regularly to a babysitter while the parents work, the sitter should also be instructed in the child's handling and positioning. Grandparents and siblings should be taught correct handling and positioning techniques. Siblings can be particularly helpful in play activities to encourage the child to move.

Positioning is one of the first components of therapy that should be considered for the child. Positioning should incorporate all aspects of the child's daily routine such as sleeping, feeding, sitting with support, bathing, carrying and transportation, and positions for play. Positioning depends on the amount of head and trunk control the child has and the motor patterns and postures he displays. Positioning and adaptive and supportive equipment should be individualized to allow the child to achieve the most control over his movement without abnormal postures in order to prevent development of back and joint deformities (Figs. 14.1, 14.2, 14.3). These positions should be both comfortable and beneficial.

When positioning a child, the therapist is aware of the influence that changes in position of the child's head and body have on abnormal muscle tone, postures, and patterns of movement. For example, the child with spastic quadriplegia may be very stiff when placed on his back; the shoulders and head pull back, the legs are straight at the knees and the feet point downward, while the elbows are flexed and the hands tightly fisted. When this child is held in a standing position, his legs may become even stiffer causing them to "scissor" (cross one over the other). When placed in the sitting position, his head and shoulders again pull back. The child cannot maintain the standing or sitting positions because his increased muscle tone interferes with balance. Positioning this child on his back avoids the problem of balance but offers him very little stimulation, as he will only be able to look at the ceiling. Since he is unable to change his position he soon becomes uncomfortable. Moreover, he is unable to bring his hands together to play with toys. A better position for this child would be lying on his side. In this position his muscles relax, his head can be forward, he can bring his hands together and watch his hands as he plays with a toy. Since the child cannot sit by himself without support, he will function better if he is seated in a pumpkin chair or corner seat tilted slightly backwards (Fig. 14.4). In this position he will have his hands free for play and be more relaxed if he is not afraid of toppling over. Placing the child in an upright position gives him more visual stimu-

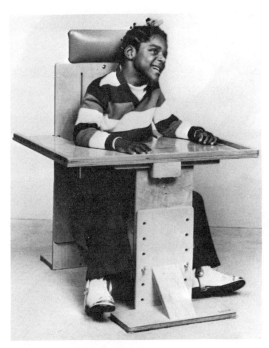

Figure 14.1. The bolster chair positions the child's legs for sitting and may help prevent the development of hip deformities. (Reprinted with permission of Kaye Products, Inc., 1010 E. Pettigrew St., Durham, NC 27701, June 1980. Bolster Chair: Model B2.)

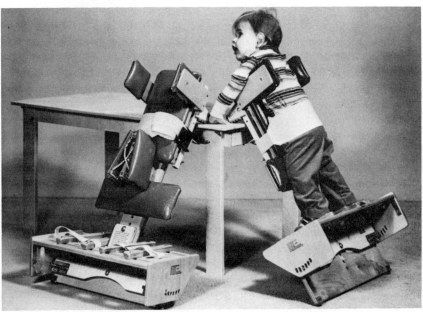

Figure 14.2. The prone stander offers the child support for standing and allows the hands to be free for play. (Reprinted with permission of Kaye Products, Inc., 1010 E. Pettigrew St., Durham, NC 27701, June 1980. Preschool Prone Stander: Model 106.)

Figure 14.3. The toilet trainer with trunk support allows the child to feel secure and relax while she is supported for toileting. (Reprinted with permission of Kaye Products, Inc., 1010 E. Pettigrew St., Durham, NC 27701, June 1980. Toilet Trainer with trunk support: Model T1.)

lation; with a lap tray he can play with toys, and he can be fed in this position. Usually the child will be more alert in the upright position than when lying on his back. The child does not stand or walk alone because of spasticity, poorly developed head and trunk control, and inadequate balancing reactions. However, he can be positioned to stand with support for short periods of time (Fig. 14.2). This will help him develop better head control and aids in the development of the hip socket during his growth period which may avoid dislocation of the hip. A child with a dislocated hip is likely to develop back deformities because his sitting base is uneven, and as a result, he tends to lean toward the dislocated hip when seated.

The child can be carried in several positions. The saddle position (Fig. 14.5) and the ring carrying position allow the child to exercise head and trunk control as much as possible and prevents the scissoring posturing of his legs. Most parents tend to carry a child who has poor head control cradled like an infant. This reduces his chance of developing head control or to visually scan the environment.

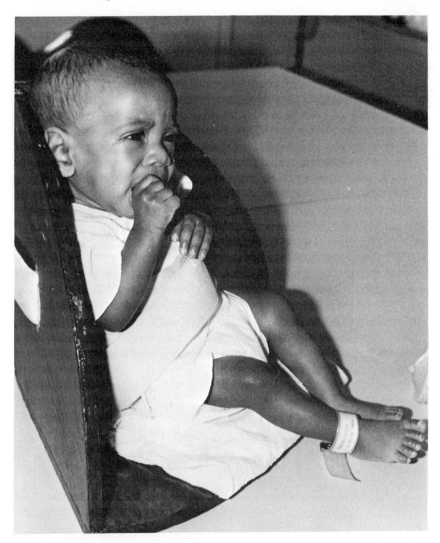

Figure 14.4. The corner chair is used to support the child for sitting and allows the hands to be free for play activities.

If the child does not have adequate head and trunk control as he grows, he will require support for transportation. A variety of chairs are available that offer support systems; the therapist should prescribe an appropriate chair for the child. The chair must fit the child correctly and also be convenient for parents to put into their car. No child should be left unattended in a chair or allowed to sit in a chair more than

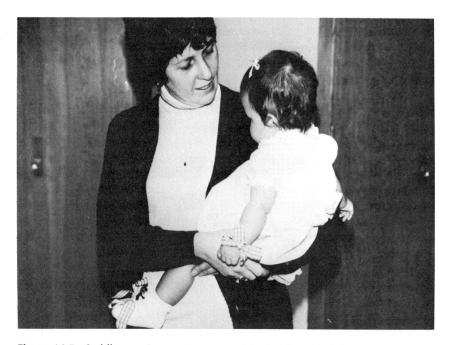

Figure 14.5. Saddle carrying position is used for holding this infant.

an hour or two. The more the child sits in the chair, the greater his chances of developing tightness in his hips and knees. One should *never* remove the foot supports of any chair and allow the child's feet to dangle. The feet have a tendency to point downward and not supporting the feet promotes contractures that will, if severe, prevent the child from wearing shoes or standing. It is important to keep changing the child's position periodically to lessen the chance of developing contractures.

The therapist should also instruct parents in the kinds of positions, toys, and activities to avoid. Among those that we would not recommend are: "W-sitting" or "reversed tailor sitting," or a position in which a child with hemiplegia sits asymmetrically towards his uninvolved side. An activity to avoid is bouncing a child in the standing position since this causes him to extend his legs more stiffly. Bouncing swings, jumpers, or baby walkers that do not encourage development of balance reactions and accentuate the tendency of the feet to point downward should not be used. These devices tend to increase stiffness in the leg muscles of the spastic child and are potentially harmful to ankle and hip joints.

Range of motion exercises in conjunction with proper handling and positioning can help prevent deformities. A child who has a neurological disorder that causes him to have abnormal muscle tone, whether it is spastic or hypotonic, will not be able to move his legs and arms normally. As a result of the lack of normal movement, the child may develop skeletal deformities such as joint contractures. A joint contracture limits the range of movement of a limb. Range of motion exercises are designed to help maintain full joint movement. Passive range of motion exercises are those in which the joint is moved by an external force without the child voluntarily contracting his own muscles to aid in the movement. Since a parent or the therapist performs the passive range of motion, this exercise does not help increase the child's strength. These exercises are beneficial in preventing deformities for cosmetic purposes, for the maintenance of functional motion in a joint such as the ankle so that the child can stand with his feet flat, and for hygienic reasons such as being able to wash under the arms, between the legs, and in the palm of a fisted hand to prevent skin problems. The specific range of motion exercises that a therapist teaches the parent should be done twice daily unless instructed otherwise. The exercises can be done during a diaper change in the morning and evening or, with the older child, before dressing him in the morning and after undressing him in the evening. Each session should not take more than fifteen minutes. By doing exercises along with a routine daily activity, they will not be forgotten or easily omitted. If the child's clothes or diapers are generally changed in the same place, such as a changing table, crib, or bed, we suggest that the therapist's instructions be posted nearby so they can be referred to easily. The therapist should instruct parents specifically in the passive range of motion exercises which their child requires so that they can perform the exercises smoothly and safely. If the exercises are done infrequently or incorrectly, they may be ineffective or even potentially harmful. The exercises should be done gently and slowly without exerting too much force or "bouncing" the joint at the end of its available range. Table 14.2 contains common "dos and don'ts" concerning passive range of motion exercises. Passive exercises only help to maintain range of motion; they do not help the child learn to move his arms and legs himself and should never be the only therapy provided to a child.

A child may have to be referred for medical consultation or management if contractures develop very early and joint motion cannot be maintained or reestablished by range of motion exercises, positioning, and active movement. The physical therapist, orthopedic surgeon, and parent should work closely together, along with the primary physician

Table 14.2. Dos and don'ts of passive range of motion exercises

"DOS"

1. DO the exercises regularly on a daily basis.
2. DO the exercises as instructed by the therapist; do not modify them.
3. DO all exercises gently and slowly.
4. DO ask the therapist to observe the exercises periodically to be sure that her instructions are being followed consistently.
5. DO have the child's range of motion reevaluated periodically by a therapist.

"DON'TS"

1. DO NOT move the arm or leg rapidly through the range of motion. This can increase resistance to movement in the spastic child.
2. DO NOT attempt to straighten the elbow or knee by pulling above and below the joint. This can increase the muscle tone and the child will pull the arm or leg into flexion.
3. DO NOT "bounce" or "spring" the joint once the end of the available range of motion is reached. Instead give a gentle prolonged stretch to the muscles for three to five seconds.
4. DO NOT pull the thumb out of a fisted hand or push up on the ball of the foot.
5. DO NOT force the motion, it is possible to overstretch a muscle or cause injury to a joint.
6. DO NOT attempt an exercise that has not been taught by the therapist or before performance is judged to be satisfactory.

to plan for the bracing, surgical, and therapeutic needs of the child. In our opinion, the most appropriate and effective therapeutic approach to the correction of motor problems should be based on the normal developmental sequence. Motor development progresses in a head-to-toe direction with head control, lying propped on the arms while on the stomach, control of trunk muscles sufficient to permit rolling and sitting (Fig. 14.6a), control of lower trunk and legs when creeping (Fig. 14.6b), and finally standing and walking (Fig. 14.7). The therapist should educate parents in normal motor developmental sequences so that they will better understand the ultimate goals of therapy.

The initial step in motor development is head control. Good head control is necessary for further motor development. The child should be discouraged from turning his head only to one side and maintaining this position most of the time. In therapy, head control, lifting the head, and holding the head midline in alignment with the body are stressed while the child is on his stomach or supported in the sitting position. It is important to practice turning the head to both sides. Balance reactions depend in part on the position of the head. Properly positioning the head and adjusting to head turning are automatic for

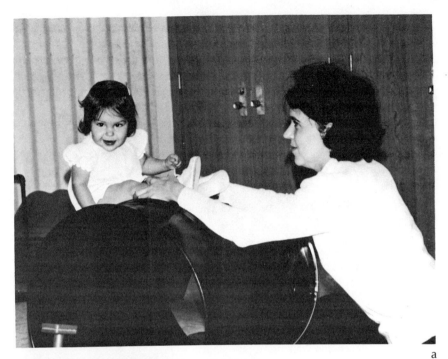

Figure 14.6. Trunk balance and weight shifting activities are used in preparation for better sitting balance and creeping on hands and knees. Balancing on and inside a large cylinder is helpful when training children to perform these activities.

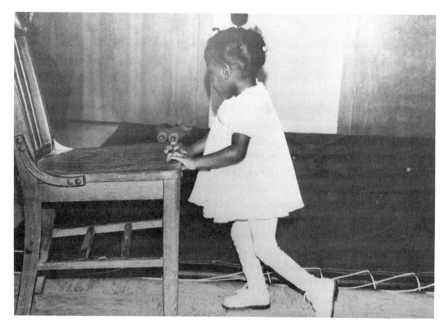

Figure 14.7. The child with more advanced motor development may use minimal support for balance and walking.

normal children but the handicapped child may need to practice these movements to elicit appropriate postural adjustments.

Motor development occurs sequentially in vertical, horizontal and diagonal planes. Head and trunk extension develop first and represent the vertical plane. Weight shifting from one arm to the other while propped on the stomach occurs in the horizontal plane. Rolling and changing the position of the trunk requires rotation along the body axis and occurs in the diagonal plane. The therapist should teach parents activities they can use to help their child initiate trunk rotation.

The neurodevelopmental approach to treatment stresses that the child be trained using the normal sequence of motor development. The child is encouraged to learn to actively move his own body by positioning him and providing him with the assistance necessary to initiate the desired movement or hold a position that maintains normal joint alignment. He is taught to develop balance and protective reactions whenever possible. By encouraging the child to actively move himself with the support from the parent or therapist, he learns through the movement experience much as the normal infant does. Therapeutic approaches that require passive movement of the child by taking his arms and legs through "patterning," such as for crawling, do not appear as beneficial or as effective because they do not offer the child the opportunity to actively move, stabilize, and balance himself and to learn to react to a movement.

Some children require more extensive therapy than that discussed. Others may only benefit from positioning and range of motion exercises because of associated mental handicaps that are so severe that movement is unlikely. Many children will require considerable therapy because of feeding problems or difficulty using hands for play (Fig. 14.8) and activities such as dressing or bathing; these activities should be included in a therapeutic intervention program.

The therapist is also responsible for recommending equipment that is appropriate for a child. This may range from an easily obtainable umbrella stroller to an elaborate wheelchair or an adaptive piece of equipment to aid a child in feeding or dressing to help him to become independent.

Figure 14.8. Fine motor activities encouraged during therapy may emphasize the use of one or both hands for play.

REFERENCES

Bleck, E. E. 1979. Orthopedic management of cerebral palsy. W.B. Saunders Co., Philadelphia, p. 266.

Bobath, B. 1967. The very early treatment of cerebral palsy. Develop. Med. Child Neurol. 9:373–390.

Bobath, B., and K. Bobath. 1975. Motor development in the different types of cerebral palsy. Heinemann Medical Books, London, p. 105.

Cooper, I. S. 1976. Living with chronic neurologic disease: A handbook for patients and family. Norton, New York, p. 318.

Cruickshank, W. M., ed. 1976. Cerebral palsy: A developmental disability, 3d ed. Syracuse University Press, Syracuse, N.Y., p. 623.

Finnie, N. R. 1975. Handling the young cerebral palsied child at home. Dutton, New York, p. 224.

Hoshell, S. H., E. Barrett, and H. Taylor. 1977. The education of motor and neurologically handicapped children. Croom Helm, London, p. 240.

Joel, G. F. 1975. So your child has cerebral palsy. University of New Mexico Press, Albuquerque, p. 53.

Levitt, S. 1977. Treatment of cerebral palsy and motor delay. Blackwell, Oxford, p. 269.

O'Reilly, D. E. 1975. Care of the cerebral palsied: Outcome of the past and needs for the future. Develop. Med. Child Neurol. 17:141–149.

Palfrey, J. S., R. C. Mervis, and J. A. Butler. 1978. New directions in the evaluation and education of handicapped children. New England Journal of Medicine 298:819–824.

Robinault, I. P., ed. 1973. Functional aids for the multiple handicapped. Harper & Row, New York, p. 233.

Rose, M. R., G. R. Clark, and M. S. Kivitz. 1977. Habilitation of the handicapped: New dimensions in programs for the developmentally disabled. University Park Press, Baltimore, p. 371.

Scherzer, A. L., V. Mike, and J. Ilson. 1976. Physical therapy as a determinant of change in the cerebral palsied infant. Pediatrics 58:47–52.

Vining, E. P. G., P. J. Accardo, J. E. Rubinstein, S. E. Farrell, and N. J. Roizen. 1976. Cerebral palsy: A pediatric developmentalist's overview. Amer. J. Dis. Child. 130:643–649.

15

BIRTH DEFECTS

A birth defect is an abnormality present in a baby at birth that is not the result of damage occurring during the birth process. A birth defect can be something obvious that does not involve the nervous system at all, such as an extra finger; it can be a defect in an internal organ such as the heart or the kidney. The more common defects that involve the nervous system are a failure of fusion of the spine and hydrocephalus, an enlargement of the head due to excess fluid in the spaces within the brain. In hydrocephalus the fluid spaces within the head may enlarge at the expense of the brain as well as increase head size. Many causes of mental retardation can also be described as birth defects, but in this chapter our discussion will be limited to myelomeningocele and hydrocephalus. Most myelomeningoceles are located in the lower part of the back in the lower thoracic or lumbosacral areas. They result in defects in bowel and bladder control and paralysis with loss of sensation in the legs and part of the trunk.

THE FAMILY PHYSICIAN

It is often the family physician who has the unhappy task of informing parents that their child has a birth defect. It is always better if the obstetrician accompanies the family doctor when the parents are told. Both parents should be present, not only because they need each other's support, but also because both parents should be involved in the critical decisions that must be made quickly if further damage to the child is to be minimized or avoided. Moreover, since these decisions must be made during a period of emotional turmoil for the family, one parent should not have to shoulder the responsibility alone.

The doctor should explain exactly what the defect is and the extent of the neurological damage. This usually can be defined after an initial examination. A family should be fully informed about which disabilities may improve with treatment and which will not. They should also be told what will happen if the baby is left untreated. Some birth defects need immediate surgical intervention. Others may be so extensive that treatment is either impossible or undesirable. A wise physician will seek the support of a specialist in neurology or neurosurgery before indicating to a family that treatment would only prolong life without improving the child's function. Even under these circumstances, the decision to prolong life or not to do so must be made by the family and not by the doctor. A physician's medical opinion should not replace a family's moral judgment. It is usually not appropriate to discuss the genetics of birth defects with the family at this time unless the parents direct questions to the doctors. The physician has an obligation to do so at a later date, however.

CASE 15.1

BJ was born with a very large defect in his back, which was diagnosed immediately as a myelomeningocele (spina bifida with protrusion of the cord). It was associated with total paralysis of his lower extremities. BJ also had an abnormally large head. The mother asked to see the baby and he was brought to her by a nurse. The mother became hysterical when she saw the defect but an intern told her not to worry, the baby would soon be dead. The family's pediatrician arrived several hours later. He found the patient's mother weeping and the father threatening to sue the obstetrician because of the back defect.

Birth defects of this type are not the result of poor medical care during pregnancy or at the time of delivery. However, the parents were justifiably upset and angry by the callous way in which the news of this tragedy was presented to them. When such a serious handicap is evident at birth, it is important that the family physician be notified immediately and that he be present to discuss the problem with the family before they are allowed to see the baby. In this instance, the pediatrician was unavailable because he was involved in another emergency. He found himself in an extremely awkward position, which was eased only by the fact that he had cared for the couple's other two children for the past five years.

He explained the nature of their child's birth defect to them, indicating that it was the result of an injury to the nervous system that occurred within the first weeks of pregnancy and not the result of an error in medical care. He then brought up the problem of care for the child, explaining alternative methods of treatment. He explained that the defect in the back was too large to correct immediately by surgery but might be correctable in the future if the baby survived. He also told the family frankly that if the child did survive, he definitely would have no function in his lower extremities and no bladder control. He might have some reflex bowel control, but that could not be assured. The pediatrician also noted that the child almost certainly had excessive fluid in his head and that this too would require surgical correction at some point in the near future. He indicated that it was his belief that when the impairment was this severe at birth a family should do nothing but sustain the patient's life. On the other hand, he said that he would call a specialist in child neurology and a neurosurgeon for an opinion and ask them to care for the child if that was the wish of the parents. He indicated that the advice he was giving was not entirely medical but represented a moral belief on his part which might differ from that of the family and that although this was a difficult time for them, decisions about extraordinary procedures were their responsibility. The physician can only give the family his opinion. The family decided to allow nature to take its course, and the child died four days later.

Comment:

One may disagree with the beliefs of the pediatrician in this instance, but they are accepted by a large segment of the medical community. Birth defects associated with *total* paralysis of the lower extremities and evidence of hydrocephalus at birth have a uniformly poor outcome. These children require multiple surgical procedures and are frequently subject to repeated infections. Many whose defects are this severe at birth have serious intellectual deficits. This is an extremely difficult decision to present to a family under the best of circumstances. It is a decision that must be made immediately and at a time when the family is least prepared emotionally to make important decisions. The family must be treated with sensitivity and feeling, and the situation is best discussed with them by someone whom they have known for a long period of time and in whose abilities they have great confidence. It is extremely important that the family be told the truth. It is equally important that the physician distinguish between what he considers to be moral and medical issues.

THE MEDICAL SPECIALIST

Many medical specialists and allied professionals are of necessity involved in the continuing care of a child with a birth defect. These include neurosurgeons, child neurologists, orthopedists, and urologists. The information the specialist should be prepared to offer the parents of a child with a neurological birth defect depends on the age of the child and the complexity of his problems.

1. Since these defects are usually obvious at birth, decisions about treatment often must be made quickly. For example, spinal defects, such as a myelomeningocele, may need to be repaired shortly after birth. In many children, the defect is not covered by skin but by a thin membrane that is easily torn or pierced, making the child very susceptible to infections of the nervous system. If the myelomeningocele is not too large, it is possible to remove it and cover the area with skin. This does not improve neurological function, but does reduce the likelihood of infection and eventually makes it possible for the child to lie on his back. If the defect is very large, a small infant may not have sufficient skin to cover the area of the defect even if it could be reduced. Under these circumstances, one must try to prevent local infections while awaiting sufficient growth to allow the neurosurgeon to cover the area with skin. As we noted in Chapter 2, if the child has no movement in any of the muscle groups of the lower extremities and has fixed contractures of the joints at the time of birth, it is highly unlikely that he will ever develop any useful function in his legs. Under these circumstances, there is considerable reluctance on the part of surgeons and neurologists to suggest that the defect be surgically repaired. This reluctance is increased by the high incidence of hydrocephalus accompanying serious cases of spina bifida.

If excessive accumulation of spinal fluid within the skull is not present at birth, it frequently occurs after repair of the myelomeningocele. Most of these children then require a second surgical procedure to treat the hydrocephalus. The use of shunts for the treatment of hydrocephalus represents one of the great advances in neurosurgery in recent decades, and it has preserved the lives and intelligence of countless children. However, it is a procedure that renders the child more liable to seizures and intracranial infections such as bacterial meningitis. Shunts often have to be revised because they become blocked or because as the child grows the tube moves out of position and needs to be lengthened. Thus, repair of a myelomeningocele in a child who has or is very likely to develop hydrocephalus generally means committing a child to long-term medical and surgical care with a strong possibility that he will require numerous surgical procedures during his lifetime. In addition to neurosurgical operations, he may also need extensive orthopedic surgery to fix his limbs in a position in which he can sit comfortably in a wheelchair or use braces to move about. He may also require surgery to allow the urine to flow freely so as to minimize the incidence of kidney infections.

Many families feel that everything possible should be done to save a child regardless of the severity of his neurological defect. This is certainly their right. In this case, the family will only need to know whether the defect is so large that it cannot be surgically repaired early in life. Clearly, a child with a very large myelomeningocele that cannot be repaired immediately must stay in the hospital a number of weeks. Before the child goes home the parents will be taught how to provide good nursing care for the child and how to keep the exposed area sterile and free of irritation. A family willing to accept this obligation certainly should expect that they will be given every aid and instruction by the nursing and medical staff of the hospital.

Most families want and should be given as much specific information as possible about the child's present level of function and possible habilitation. They will want to know the level of the defect and its size; whether there is any movement or sensation of pain below the defect; if the child's joints are immobile and therefore will eventually require surgery; and if there are other deformities. They should be particularly interested in whether the child has hydrocephalus and how severe it is. If it is very severe, it is unlikely that the child will ever be of normal intelligence even if the hydrocephalus is treated early in life. Small defects in the lower portion of the back often can be repaired easily, and residual problems are frequently limited to bowel and bladder control. There is always a possibility that hydrocephalus in these children will develop later in infancy or in childhood, but the chances are relatively small. If the hydrocephalus is then treated promptly, the child may develop normal intelligence.

Given this type of information, a family can discuss the problem intelligently with the specialist before making a decision about treatment of their child's birth defect in the first days or weeks of life. However, many families prefer *not* to be given information and instead to leave all decisions in the hands of the medical staff. While this may seem reasonable—because the medical and surgical problems involved are complex and the family is understandably distraught—it is not advisable. If children with birth defects are to survive, be comfortable, and lead reasonably happy or productive lives, they must have the continuing attention of their families. It is therefore exceedingly important that shortly after birth the family understand the nature of the defect and its implications for the child's future. In this way they can help in the decision-making process and begin to recognize the complex problems they must face as a family. During this period they may need and should be offered counseling by experienced professionals on the staff of the hospital to assist them in making responsible decisions that they will be able to live with in the future. In our opinion, professional personnel do a disservice to the family if they relieve them of any responsibility for making decisions at this early critical period.

2. A child with myelomeningocele will need the help of many professionals throughout childhood and into early adult life. It is desirable that families seek multidisciplinary clinics whose staff includes medical specialists, physical therapists, psychologists, and social workers to provide the expertise necessary for the care of these children. If such clinics are not available, families should be told that consultations with one or more medical professionals may be required at some time. These referrals will be made as required by the family physician or by a specialist in neurology or neurosurgery. Physical therapy must be started early in life to minimize joint contractures, to strengthen residual muscle activity, and to minimize abnormal curvatures of the spine. If hydrocephalus is present, occupational therapy may be needed in order to improve fine motor skills and visual motor coordination of the hands (these problems are discussed in detail in Chapter 14).

A child with a birth defect will also need to be examined periodically for urinary tract infections. The physician's goal is to prevent or minimize these infections. Children with spina bifida may dribble urine constantly, resulting in skin irritation and repeated infections of the skin of the thighs and buttocks. However, the fact that this occurs may be symptomatic of a condition known as overflow incontinence. In this disorder, a large residue of urine remains in the child's bladder and is a constant source of infection. Furthermore, the tubes from the kidney to the bladder may dilate and allow infected urine to back up into the kidney. If this condition is left untreated, permanent kidney damage can result. Parents must be trained to help the child void by manual compression of his bladder through the abdominal wall or by catheterization. Later many children can learn to perform these maneuvers for themselves.

In addition to frequent examinations of the urine to check for infection, periodic x-ray examinations of the kidneys are needed to determine whether the structure and function of the kidney is normal. If repeated infections occur or there is dilatation of the tubes that lead from the kidney to the bladder, reconstructive urinary tract surgery may be necessary. Parents of a child with a birth defect should make sure that the child's condition is reviewed regularly by a urologist.

Children with myelomeningocele may be severely impaired physically but still have normal intelligence. Their mental abilities should be assessed by appropriate psychological tests when they are ready to enter preschool or kindergarten. There is no reason why a child with a myelomeningocele cannot be an independent, productive member of society earning his own living if he has reasonably normal intellect and good use of his upper extremities. If these children are normal or bright and their hands are unaffected, the only limitation that they have educationally is their ability to get to and from school and from class to class within the school. It is now the legal obligation of the school system to provide for the transportation of these children and to arrange school facilities so that they can get to them by wheelchair, if necessary. Often schools will suggest that the child be taught by a home tutor rather than attend a regular school. In a sense, this is easier for the child and often less expensive for the school. However, if parents are given this choice for their child, they should bear in mind that the child must eventually learn to move about in society, even if in a wheelchair. The child must therefore learn those skills that will ultimately lead him to independence, and he must associate with children and adults who are not physically handicapped. The earlier he is taught the skills he will need, and the sooner he begins to associate with others, the better it will be for him. It may be necessary to enlist the aid of physicians and other professionals who are caring for the child to persuade the school that he has a right to a normal education. Frequently social workers and educational therapists as well as psychologists who work closely with birth defect clinics are aware of the resources in the community that will allow this type of educational and social integration to occur. Parents should be in contact with them and insist that they make every effort on their child's behalf.

3. The pattern of inheritance of birth defects is not known. If a family has one child with a myelomeningocele, there is approximately a 5 percent chance that they will have another child with a similar defect. Recent advances in the examination of unborn infants (either by direct visual techniques, such as the use of the fetoscope or by taking pictures of the fetus by ultrasound, or biochemical measurements of a substance called alpha-fetoprotein) have made it possible to diagnose a myelomeningocele or similar defect *in utero* and to abort the child if that is the family's wish. Families who have one child with a birth defect could avail themselves of these procedures.

THE PSYCHOLOGIST

A child with a myelomeningocele who has no impairment in the use of his upper extremities can be evaluated with the same tests used for other children of the same age (See Chapter 5). The psychologist may also help with counseling and behavioral problems, as described in Chapter 6.

THE EDUCATOR

The education of a mentally normal child with a motor handicap is discussed in Chapter 14.

THE PHYSICAL AND OCCUPATIONAL THERAPIST

Children with spina bifida and myelomeningocele need attention from many specialists. Physical and occupational therapists should be regularly involved in the management of the child with myelomeningocele as a part of the health care team. The PT/OT management of the child with myelomeningocele will vary according to four major factors: (1) the extent of damage to the spinal cord as a result of the birth defect; (2) the absence or presence of hydrocephalus; (3) the child's age at the time of referral to physical and/or occupational therapy; and (4) the child's home environment. Therapists should be aware of problems that result from the myelomeningocele defect and be able to evaluate the extent of functional impairment that can be associated with these disorders. Impairment could include: (1) loss of feeling in the skin or decreased sensation in the trunk and legs, which can vary depending on which nerves are damaged; (2) limitations in joint range of motion; (3) back and postural deformities due to malformation of the spine, and weak trunk and hip musculature that cannot support the spine as the child grows; (4) delayed development that can be the result of weakness in the trunk and legs or of brain damage sometimes associated with the myelomeningocele; (5) paralysis of muscles in trunk and legs; (6) lack of bowel and bladder control; and (7) hydrocephalus, which may cause mental retardation or difficulty in learning.

Early management

The physical and occupational therapist's involvement with a child's care should begin during the first week of life while the child is still in the hospital. At this time the therapist's concern and efforts are primarily aimed at improving or maintaining the joint motion the child has in his legs and feet, positioning the child appropriately, and teaching the nursing staff and parents how to perform these activities.

The physical therapist is able to evaluate how strong the infant's legs are by observing his spontaneous movements when he is awake and active or angry. The therapist may also test muscle strength by feeling how the muscles move under the skin as they contract. Joint motion is evaluated by measuring each joint with a tool called a goniometer. Sensation in the trunk and legs is usually evaluated by watching the baby's response to pinprick. This information, along with muscle testing results, helps to determine what parts of the spinal cord are functioning. The therapist may also evaluate the infant's ability to suck and swallow, his posture when on his stomach and back (if the baby can be placed on his back at this time), how well he lifts and turns his head, and his ability to attend to sounds and to follow faces or brightly colored objects with his eyes. A check of these functions should be included in the documentation of a child's development.

Many children with myelomeningocele lie with their legs in a frog-leg position due to muscle weakness or paralysis (Fig. 15.1). The child may not have the same ability to move out of this position as an infant with normal strength who would be able to kick and move his legs actively. The child with myelomeningocele is at risk for developing joint contractures (shortening of the muscles) and joint deformities if he does not move his legs (or have someone move his legs for him), or if he is not positioned correctly. While the infant is in the hospital, the therapist will begin joint range of motion exercises of the legs (Fig. 15.2) and work with the nurse to teach her how to position the infant for sleeping, feeding, and carrying (Fig. 15.3).

The therapist is also responsible for instructing the parents or caregiver in exercises, positioning, and correct handling of the infant. In most hospitals the parents are encouraged to participate in the child's care from the beginning. Gradually parents are taught range of motion exercises, correct positioning and handling techniques, bowel and bladder care, skin care, special feeding techniques if needed, and splint wearing schedules. This is a great deal of information for parents to learn. Therefore, it must be taught to parents over a period of several days, not one hour before the baby is to be discharged from the hospital. When the therapists teach an exercise or activity, they should demonstrate and explain the exercise first, allow the parents to practice it and then to demonstrate it to the therapist. The therapist should give the parent a brief written description or illustration of the exercise along with written instructions indicating how often the exercise is to be done at home. It is most important that parents understand the purpose of the exercises and that they know how to perform them safely and easily with their infant.

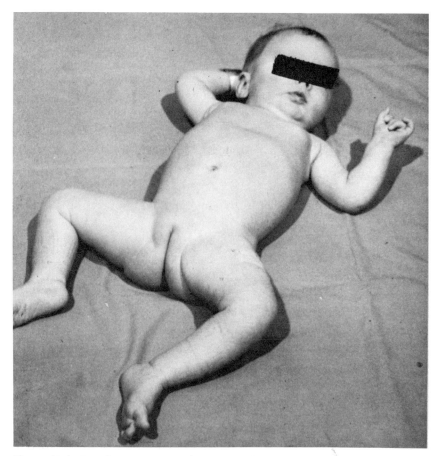

Figure 15.1. Frog-leg posturing is demonstrated, indicating a severe decrease in tone at the hips.

When an infant is discharged from the hospital, parents become the baby's primary therapists. The physical therapist and occupational therapist will either follow the baby through a facility-based outpatient clinic, a home-based program, or a combination of the two. The therapist will want to see the baby frequently during the first year to monitor his motor development, strength, and range of motion, and to instruct parents in changes in the baby's therapy program as his needs change. If a child is followed regularly in a therapy program, this will be done routinely. If the family lives too far from a facility or home-based program to see a therapist regularly, they should return

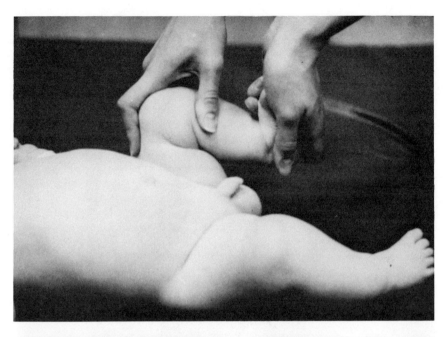

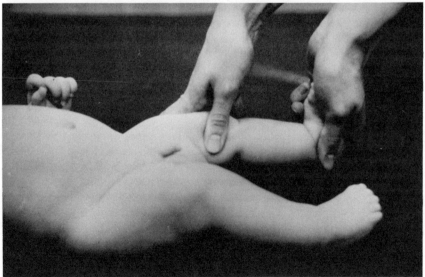

Figure 15.2. Range of motion is done to the legs to stretch the heel cords.

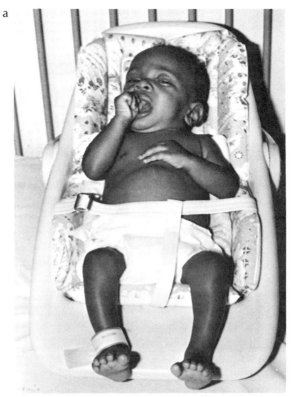

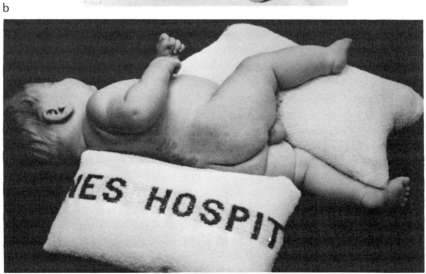

Figure 15.3. (a) The pumpkin seat or infant carrier may be used to position the infant for feeding. (b) The sidelying position may be used for sleeping.

when the child is one, three, six, nine, twelve, eighteen, and twenty-four months of age. This is the period when the infant develops most rapidly. Since children with myelomeningocele vary in their degree of disability and developmental abilities, only general principles can be covered in this text. The therapist should be able to determine the best individualized program for each child.

Skin care

Skin care is a major component of a therapy program for a child with myelomeningocele because the skin serves to help control body temperature and guard against infection. The condition of the skin can directly influence whether a child can wear his braces and maintain a good activity level. The skin is considered to be in good condition when it is free from open sores, scrapes, red areas, blisters, or bruises. The child who has chronically open sores may develop decubitus ulcers and will face repeated hospitalizations to heal the sores. He may eventually require plastic surgery to close the ulcers. The old adage "An ounce of prevention is worth a pound of cure" certainly applies to skin care.

The first step in prevention of skin breakdown is to be aware of some of the factors that influence the integrity of the skin. Bacteria multiply rapidly on moist areas of the skin, particularly if they are wet with urine. The skin becomes prone to irritation and eventually breaks down. The skin should be kept dry and clean to avoid bacterial growth. Prolonged pressure on an area of the skin such as the buttocks from sitting for long periods can cause the skin and tissues underneath to break down. The constant pressure reduces the blood supply to the area. When this occurs, the cells may die and a decubitus ulcer can start to form. This is apt to occur in the child with myelomeningocele who sits most of the day in a wheelchair unless he is taught to shift his weight frequently or to do wheelchair pushups. The child who does not have intact skin sensation over the buttocks cannot feel discomfort. The normal person feels uncomfortable when blood supply is cut off and changes position, allowing renewed blood flow to compressed areas. The child with decreased sensation must be trained to remind himself to change position, since he does not feel uncomfortable. Contact with the floor (from scooting or crawling) and the chafing of braces, splints, or shoes can also cause the skin to break down.

The therapist can teach a child weight shifting and wheelchair pushups. Equipment can be provided that allows better movement on the floor and avoids endangering the skin. The therapist can also teach proper transfer techniques and help the child learn to move about without scooting or dragging his knees or buttocks on the floor. To prevent the child from developing pressure sores and skin breakdown, parents should observe the following rules:

1. Know the areas of the skin where skin breakdown is most likely to occur and check these areas daily (Fig. 15.4).

2. Keep the skin clean.

3. Keep the skin dry.

4. Fit shoes and braces properly and build skin tolerance gradually.

5. Dress the child who crawls on the floor in leotard tights or long pants to protect the legs.

6. Do not let the child scoot on the floor, as this causes skin friction. Use a scooterboard to protect the skin on the legs (Fig. 15.5).

7. Teach the child skin care early, and to use a hand mirror to check skin on bottom and back of legs.

8. Use only cushions that are designed for wheelchairs. Pillows not

Figure 15.4. Areas which are prone to breakdown on the skin are circled.

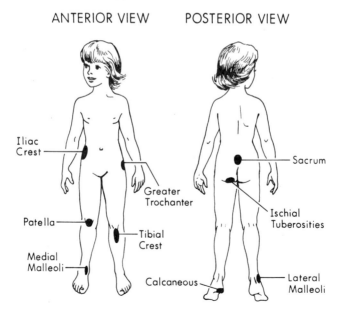

ANTERIOR VIEW   POSTERIOR VIEW

AREAS PRONE TO PRESSURE SORES

Figure 15.5 The scooterboard can prevent floor and rug burns on the knees of the child with poor sensation in the legs.

designed for this purpose do not offer sufficient protection against skin breakdown.

9. Children confined to a wheelchair should be reminded to do pushups each hour.

When a child receives new braces or new shoes, use the schedule for building skin tolerance recommended by the child's physical therapist or orthotist. This schedule will gradually increase the time the child wears his braces and will help avoid red areas and blisters due to skin contact with the brace or shoe. Such a schedule will gradually increase the number of hours that new braces or shoes can be worn. Shoes should be put on carefully. Open-toe shoes are best for infants, toddlers, and nonambulatory children. Lace-to-toe surgical boots are better for children with myelomeningocele than standard high-top shoes because it is easier to check the toes to ensure that they are straight when the shoes are put on. If shoes or braces rub, cause blisters, or create a red area that does not fade within fifteen minutes after the appliance is removed, the brace or shoe must not be padded with gauze or moleskin, as this will cause further irritation to the area. Contact the physical therapist, orthotist, or physician for instructions. A good rule to use whenever possible is "remove the source of irritation immediately and avoid pressure to or contact with the area until it has healed."

Maintaining joint range of motion

Joint range of motion is the number of degrees of movement that a joint can pass through. For example, the normal knee joint can extend or straighten to 0° and can flex or bend to 145°. Our muscles act to pull on the bones and joints to move them through the planes of motion. When a child has weakness or paralysis, the muscles are not able to move the joint through the full motion, or the muscles of the joint may not balance each other. When this occurs, the range of joint motion decreases and contractures develop. Contractures are fixed joint deformities that develop when a joint is not moved through its full range, causing the muscles and tendons to shorten.

It is important to maintain the range of motion in the infant's legs so that he will be able to wear shoes and braces and to stand or walk. Maintaining good range of motion will also make it easier for the child to dress himself, maintain good hygiene, and have a better physical appearance. Some children with myelomeningocele are born with joint contractures or deformities such as dislocated hips or club feet. When a contracture cannot be corrected by range of motion exercises, splinting may be done by the therapist or the orthotist to better position the joint. Surgery is sometimes necessary if splinting and range of motion exercises cannot correct the problem.

A child with myelomeningocele is at a greater risk for development of contractures in his legs if he is nonambulatory and sits in a chair or remains in bed most of the day. These children require a daily program of passive joint range of motion exercises to prevent them from losing joint motion. The child who is in a wheelchair most of the day is likely to develop tightness in the hip and knee musculature. This can be avoided by exercises performed while he is lying on his stomach. Several brief periods of exercise daily are needed to stretch out the hip and knee flexor muscles. At five to six years of age, a child of normal intelligence can begin to perform some range of motion exercises for his legs by himself. The therapist will be able to determine when a child is ready to learn these exercises.

Posture

Posture refers to the alignment of the body's parts. Alignment when the body is motionless is called "static posture"; when the body is in motion it is called "dynamic posture." Postural problems can occur as a result of decreased joint range of motion, which in turn results in poor alignment of hips, knees, and ankles with the spine. Children with myelomeningocele often have deformities of the back due to malformation of sections of the spinal column, the vertebrae. Scoliosis is a term used to describe back deformities in which there is curvature of the spine and the spinal column is rotated. A child who has scoliosis may need to wear a body jacket or brace to correct the problem or to prevent worsening of the condition. Physical therapists work with children with postural problems and scoliosis to correct minor problems by exercise. They also teach positioning and good posture habits to keep postural faults from progressing. Therapists will instruct children who wear body jackets or braces in posture and breathing exercises in order to maintain good chest expansion. They will also teach children how to use the jacket or brace to achieve the best possible posture.

The physical therapist is aware that children with myelomeningocele are at risk for development of postural problems and will evaluate a child's posture regularly. An orthopedic surgeon should also examine the child for back deformities and check x-rays of the spine and hips once or twice each year to compare the child's spine and hip placement with that seen on previous films. If a child's back deformities are severe and cannot be stabilized or corrected by bracing and exercise, the orthopedic surgeon may attempt to treat the problem surgically. While the child is hospitalized, he should be seen by the physical therapist, who will help him maintain chest expansion so that he can move sufficient amounts of air in and out of his lungs. The therapist should also teach the child how to increase the strength in his arms. She will insure continued range of motion in the joints by positioning and gentle exercise.

Aids for standing and walking

Most children with myelomeningocele who have good head control and arm strength and are not severely impaired mentally have the potential to stand and walk with bracing. Standing is important for the child with myelomeningocele because it allows him to be at the same eye level when he interacts with other children his age. The increase in physical activity associated with standing helps improve circulation. Moreover, it is easier for the child to breathe when standing. Standing also strengthens the bones in the legs and helps bowel and bladder function.

Bracing usually begins sometime before eighteen months of age for children with myelomeningocele. The type of brace a child needs will depend on the muscle function and strength he has in his legs. The

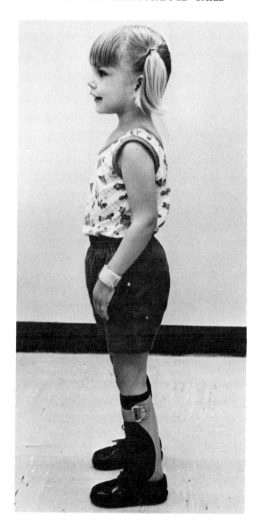

Figure 15.6. The polypropylene ankle-foot orthosis maintains good foot positioning for standing and walking. It is shown here on a child with hemiplegia.

child who has strong hip and knee muscles but weak ankle and foot muscles may only need a polypropylene insert orthosis to stabilize the ankle and foot for walking (Fig. 15.6). Children with less function in the legs may require braces that stabilize the feet, ankles, and knees. Children with complete paralysis of the legs will require bracing above the hips. These children can begin a standing program between twelve and eighteen months of age when they start demonstrating a desire to stand by trying to pull up. They are taught to use a standing

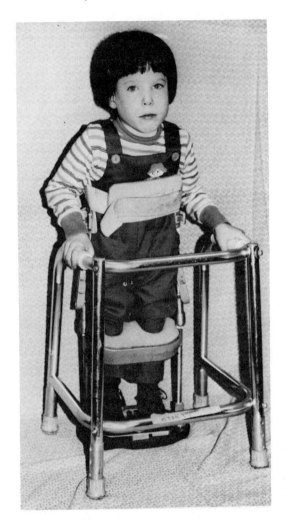

Figure 15.7. The child with complete paralysis in the legs requires more extensive bracing such as the parapodium.

brace or parapodium at first and later progress to long leg braces with a pelvic band support (Fig. 15.7) to steady the lower part of their trunk.

The physical therapist reports the child's achievements in motor development, strength, and range of motion to the physician and makes recommendations for bracing. When a child receives braces, he will begin gait training with his physical therapist. The program may extend over several years and should include how to remove and put on braces, how to use a walker or crutches with the braces, safety precautions, how to check skin condition, how to sit down and stand up from a chair, the correct method of falling to and getting up from the floor, and how to walk outdoors on rough ground, ramps, curbs, and stairs.

Children with greater loss of leg function may choose a wheelchair in preference to walking with braces when they are teenagers or young adults because they can move about faster and easier in a chair. This choice should be made in consultation with their physician and their therapist. It is not always an undesirable choice.

Physical therapy exercise programs for the child with myelomeningocele

Children with myelomeningocele require individually designed physical therapy exercise programs to meet their specific needs and problems. When planning an exercise program for a child, the physical therapist takes into consideration the child's strength and functional abilities along with his mental and emotional status. Obviously, the child with severe mental retardation and total paralysis of the legs will have a program much different from that of the child with normal intelligence and only minimal weakness. The physical therapy exercises will be upgraded as the child grows stronger and develops intellectually. Exercise programs in physical therapy should include strengthening, endurance, range of motion, balance, coordination, and adaptive functions.

Initially an exercise program will concentrate on maintaining or improving joint motion in the hips, knees, and feet; maintaining good trunk alignment; and development of gross and fine motor skills. During the first year the parents will primarily be responsible for administering the therapy program under the therapist's guidance (see the section on early management). The main goal of therapy in the first two years is to prepare the child for standing, which may be initiated anywhere between twelve and twenty-four months of age. The developmental program may be similar to that discussed for head and trunk control, and fine and gross motor development in the chapter on cerebral palsy. Most of the exercises can be incorporated into a parent's daily activities.

Between two and four years of age, the preschool child needs more strengthening exercises. At this age children are fairly cooperative in participating in exercise programs, but it is still the responsibility of the physical therapist and the parents to see that the exercises are performed properly. Exercises for strengthening the trunk, arm, and leg muscles should be incorporated into play and into activities of daily living such as dressing. However, the program presented at this age should include a schedule of rewards that are carefully selected and known to be reinforcing to the child to keep him interested and involved in the exercises. Some children will begin to cooperate with more formal therapeutic exercises between three and four years of age. Even if the child cooperates in formal exercises, he should still have play activities and rewards to maintain his motivation.

During the early school years the child is gradually taught to perform exercises for strengthening and range of motion by himself. By the time the child is ten to twelve years old, he may be ready to accept responsibility for most of his exercises with minimal supervision from the therapist or the parents. This independence should be encouraged, as it instills a sense of self-confidence and responsibility in the child.

Exercise programs for children with myelomeningocele concentrate on strengthening arm and trunk musculature. It is important to develop good strength in the arms and trunk because the paraplegic child will either require crutches for walking or will need to use his arms for moving about in a wheelchair. Strong arms are required for the performance of transfers to and from the wheelchair to bed, the floor, the toilet, and the bathtub. Before the child becomes a teenager, endurance exercises should be added to his exercise program. Swimming and walking, even with the aid of an appliance, are both good endurance exercises. Exercise periods should build up to at least one-half hour three times per week over several years. This will not only be beneficial to the child's general physical fitness and health; it should also increase his ability to use his arm, trunk, and leg musculature for longer periods of time before he becomes tired. As a result, he will function better in social and vocational activities.

Balance and coordination exercises are continued into adolescence and adult life. The child who is limited to a wheelchair may also need to be taught balance and coordination exercises, since he will be apprehensive of any movement if he is afraid that he will lose his balance. Many children who have weak lower trunk muscles will either have to use their hands to support themselves when sitting or wear a body jacket for trunk support. These children need to learn how to use their head, upper trunk, and arms to help them balance. The better the child's balance and coordination, the greater will be his ability to dress, transfer, bathe, and walk. He will also be less fearful of being moved or of attempting to move himself.

Equipment for the child with spina bifida

Braces, modified shoes, trunk supports, crutches, walkers, canes, and wheelchairs should never be ordered without a physician's prescription stating the exact specifications for the equipment. The family physician should consult with the orthopedic surgeon, child neurologist, or physical therapist before he prescribes equipment for a child. Prescriptions for braces and shoes should be taken to a registered orthotist. He will make sure that these items are properly fitted, function appropriately, and are constructed to minimize the chances of irritation. The therapist, as noted earlier, should begin a program to increase skin tolerance to new shoes and/or braces. She should also teach the child and family how to put on, take off, and use these items correctly.

The wheelchair is an often abused and misused piece of equipment. Because wheelchairs are very expensive, many families must look to insurance or charitable organizations for help with their purchase. Well-meaning charities often have "loaner" chairs available or will donate a large chair so a child can "grow into it" and use it a few years longer. This is exactly what the child will do—he will "grow into" the chair and conform to its shape, making him prone to all sorts of back deformities. An oversized wheelchair cannot properly support the child; it allows him to slump over. If a child must be in a wheelchair most of the day, it is very important that the wheelchair fit the child. Parents should *never* buy or accept a chair with the idea that the child will grow into it. The physical or occupational therapist should measure the child and order a wheelchair just his size. The wheelchair should fit the child correctly and support his back and feet. The arms of the chair should be at a height that will allow his arms to rest at a comfortable 90° angle at the elbows without raising his shoulders. The chair should be narrow enough to allow the child to reach the wheels to propel the chair and to reach the brakes. All wheelchairs should have a solid seat with a seat cushion that is properly fitted, a seat belt for safety, removable arm rests, and foot rests that swing out. Foot rests should never be permanently removed from a wheelchair. The child should always have his feet supported. Parents must be sure that their child's wheelchair is well maintained and serviced by the dealer, the therapist, or themselves after they have been instructed in its care. This should prolong the life of the chair and prevent breakdown of parts.

The infant or preschool child should not be allowed to drag his legs on the floor as he pulls himself with his arms. This is not safe for the

Figure 15.8. The Row Car, a type of caster cart, provides a safe means of recreational transportation for the child with spina bifida.

child's skin. A scooterboard can be used for moving about in the home (Fig. 15.5). The scooterboard is a good piece of equipment for several reasons: (1) it offers protection to the skin on the child's legs; (2) it provides the child with an effective means of mobility in the home; (3) it promotes the development of arm and trunk strength and improves balance; and (4) it is inexpensive and simple to build.

A caster cart can also provide the child a means for mobility while protecting the legs and encouraging the use of the arms (Fig. 15.8). Equipment and toys available commercially are listed in Table 15.1 categorized according to strengthening, balance, and coordination benefits.

Any special equipment for the child with spina bifida should be carefully evaluated by the physical or occupational therapist to determine the appropriateness and safety of the equipment. The toys listed in this section are easily available but are not necessarily appropriate for all children.

The child manipulates objects and explores the world in a variety of positions. However, sitting is usually the preferred position because when the child is upright with his toys in front of him, he can most easily engage in activities involving eye-hand coordination. It is

Table 15.1. Equipment and toys for the physically handicapped child

Equipment/Toy	Dealer/Manufacturer	Area strengthened			Balance coordination
		Arms	Trunk	Legs	
Row Car	Lossing	x	x		x
Irish Mail		x	x		x
Green Machine	Mattel	x	x		x
Caster cart	Homemade	x	x		x
Scooterboard	Homemade	x	x	x (if sitting)	x
Happy Hippo Scooter	J.C. Penney	x	x	x (if sitting)	x
Push Cars	Assorted		x	x	x
MiniWheels	Louis Marx Co.		x	x	x
Tricycle	Assorted		x	x	x
Tike Slide	Rotadyne	x	x		x
Wooden Doll Carriage	Community Playthings		x	x	x

more difficult for a child to use his hands for these manipulations if he is lying on his back, since he must raise his arms in order to see the toy that is in his hands. When the child is on his stomach, he shifts his weight onto one arm and then has only one arm free to reach for and play with objects.

To allow the child the greatest freedom to manipulate objects in the sitting position, he should not have to concentrate on maintaining his balance. He should be propped up with pillows, a cutout seat, a corner chair, or a cart on wheels, so that his position is maintained—and his mind and hands are free for play. If the child stands in braces, he should be propped securely against the table for play.

If a child lacks head control, a supported sitting position, such as in a pumpkin seat or lying on his side with hands together, may allow him to manipulate objects more easily. For the child with impaired mentation, play activities should reflect his functional capabilities rather than his age. Activities will need to be simplified to allow the child to enjoy the experience.

Activities of daily living

A child's potential for independent motor activity is influenced by the extent of the weakness in his legs and trunk muscles, the strength in his arms, his intelligence, and his motivation. Parents have no control over the degree of spinal cord impairment, but they can encourage their child to engage in exercises that will increase his muscle strength. Parents can also help set reasonable goals for the activities of daily living and influence their child's motivation to achieve these goals. They can motivate their child by expecting him to perform and by consistently reinforcing his efforts and achievements. An over-protective attitude may limit the child's future accomplishments. Consultation with a physical or occupational therapist often is essential to ensure realistic goal setting for the child. The goals must be outlined carefully with aims that can be achieved at each stage of the child's training so that he can be assured of some successes. A child's motivation may be destroyed if he experiences failure repeatedly.

The young child should be required to cooperate in dressing, if he is at all able, as this helps set the stage for later independence. The average infant at ten to eleven months will help pull his arms out of his shirt. Soon afterward, he will hold out his arms for his shirt when he is being dressed. At fifteen months he can pull off his hat or his socks. By two years he will be undressing himself if he is helped with buttons or snaps. By three years the normal child can dress and undress independently and will need help only with buttons and zippers. Although the handicapped child may not be dressing himself completely by three years of age, the sequence in which he learns to dress is the same: first he helps with undressing, later he helps to put his clothes on. Children learn to undress themselves before they learn to dress themselves completely. Fine motor skills such as buttoning and tying laces are acquired last.

There are several points that must be remembered when teaching the young child with spina bifida to dress himself. His lack of sensation requires that clothes and shoes be fitted carefully. Clothes should not press tightly against his skin. One of the most frequent errors made in dressing a child with spina bifida is to accidentally allow his toes to remain bent inside his shoe. This can lead to poor circulation and eventual skin breakdown. The child's balance also must be considered. If a child has poor control of his trunk muscles, he will probably not be steady enough to free both arms for dressing. Parents must be sure their child's balance is secure by supporting his body with a bed-seat cushion or by having him sit on a corner of the bed with his back against the wall for support. A natural place for a child to begin to learn to dress himself is on the bed because he can lie down to put on his pants. Shirts, tops, or dresses can be put on in bed or in a chair. A child whose intelligence is impaired will need simplified methods, but the child's success will be influenced by his parents' efforts, expectations, and consistency. If parents dress their child three or four days a week because on these mornings they are pressed for time and cannot wait for him to dress himself, a routine will be difficult to establish. A more reasonable approach would be to start teaching the child to undress at night and to begin to learn to put on his nightclothes when the family is not hurried. Parents have more time at night to allow the child to practice with supervision and nightclothes are generally loose fitting and easier to remove and put on than other clothes.

Parents may find it helpful to consult with their child's physician or therapist to be sure that the child is ready to begin learning to dress himself. The following is an example of the kind of help a child's occupational therapist should make available to parents. She should demonstrate dressing techniques to the family with the child and then allow them to practice while the therapist watches and is available to suggest changes or answer questions. This can be done on a number of occasions until the parents feel they are able to carry out the task comfortably without supervision. It is important that a consistent method of dressing be used. The child should have the opportunity to practice dressing at the same time each day. Parents should encourage a successful learning experience and reinforce the child's achievements. It is easy for the child to become frustrated when he tries to learn self-care tasks. Although the child's frustration should be acknowledged and respected, expectations for reasonable performance should not be altered merely because the task is difficult, as long as parents are sure that they have set reasonable goals in consultation with the child's therapist and physician.

Bathing and personal hygiene

Bathing and personal hygiene are other areas of self-care that can be affected by the neurological impairment associated with spina bifida. Access to the bathroom is essential. However, access can be limited if the bathroom does not accommodate a wheelchair. If the wheelchair cannot fit through the narrow door, an "adjusto-width" crank, which temporarily decreases the width of the chair, can be purchased at a medical rental facility. Minor changes in the position of the drain pipes may give the child the extra space needed to allow a wheelchair to fit under the wash basin. If the wash basin or sink is not accessible, a child who must remain in a wheelchair can use a wheelchair tray with a basin of water, a mirror, and other hygienic paraphernalia (Fig. 15.9).

The child with poor mobility may be unable to use the wash basin or get in and out of a tub. As a child gets older and heavier, it becomes more difficult for him to lift his weight safely, and it is also

Figure 15.9. When the sink is not accessible to the child in a wheelchair, he can use a lap tray with a basin of water to do hygiene activities.

more dangerous for parents to lift him. Once positioned at the sink or in the tub, the child's poor balance can result in further problems. The child must be transferred in and out of the tub without injury to himself or to his parents. Since many children lack sensation in their legs and lower trunk, the water temperature must be checked before the child is placed in the bath. To prevent back injury, family members should learn how to lift the child properly and should avoid lifting a child who is too heavy. A therapist can review good lifting techniques with parents and teach them to lift and transfer their child correctly.

Fortunately, there are many kinds of adaptive equipment available today that make bathing and personal hygiene considerably easier for the child. Although most specially ordered equipment is expensive, many items can now be purchased in department stores or built at home. The child who is unable to walk may need only a safety bar at the edge of the tub or on the wall at the side of the tub for safe entrance and exit (Fig. 15.10). The wheelchair-bound child who has very strong arms, good judgment, and motivation may be able to transfer himself from the wheelchair into the tub and back out again using a forward transfer method if there is a bathtub bench. Most children can learn to take a bath independently if they use specially adapted equip-

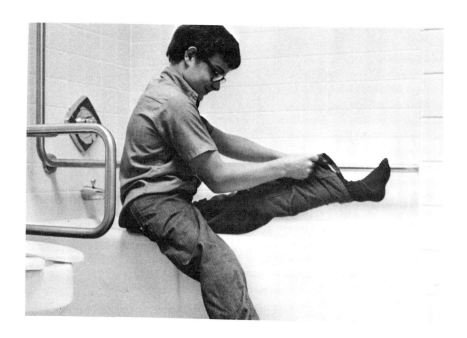

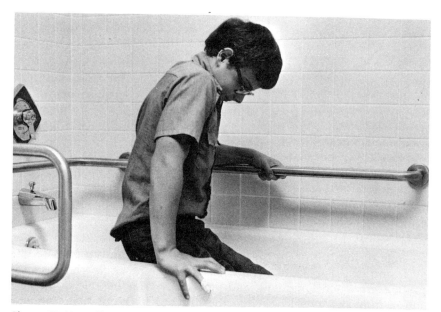

Figure 15.10. Following transfer to the side of the bathtub, this child lifts his legs over the edge and lowers himself into the tub.

ment and proper transfer techniques. The necessary adaptive equipment includes a bathtub bench, a long-handled shower hose, safety bars, and a long-handled bathtub brush. An additional safety belt or safety bar is necessary for the child with poor balance. This eliminates heavy lifting and at the same time allows the child privacy and freedom in the bath. However, the child with impaired judgment will continue to need assistance. A child's abilities and needs should be reevaluated periodically by his therapist and his physician so that those pieces of equipment most suitable for him can be chosen. However, as is the case with dressing, the hard work is done at home and requires consistency of method and positive reinforcement.

Transfers

The ability to move about is essential to a person's independence. If an individual is confined to a wheelchair, independent mobility is still possible. An appropriately equipped wheelchair with a transfer board and a safety belt along with instruction and supervised practice in appropriate transfer techniques can make movement out of the chair possible. If a child learns the appropriate transfer techniques, he can transfer safely into and out of bed (Fig. 15.11), tub, toilet, chair, or car, often without the help of other family members. In addition to following basic safety precautions (locking brakes of the wheelchair and transferring to a stable surface), the child with myelomeningocele

Figure 15.11. (a) The armrest is removed in preparation for a transfer. (b) The sliding board transfer is used from wheelchair to the bed. (c) The child slides along the board to bed.

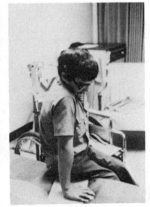

a   b   c

must avoid bumping or scraping his legs during these moves. It is important that the child be taught how to do wheelchair arm pushups before he tries transfers. A sideways transfer involves positioning the wheelchair at an angle, removing the arm rest, using a transfer board, lifting the legs while the child pushes up with his arms. The child who has had a great deal of practice may not need a transfer board. The child can transfer to the floor via graduated stairs or a ramp from the wheelchair. Many highly motivated children, given sufficient opportunity to practice, eventually learn to get onto the floor without the use of equipment. This brief description of transfers is not meant to instruct parents or children in transfers, but to indicate that techniques are available that can increase a child's independence. Therapists should teach each child the transfer techniques most appropriate to his needs.

Passive transfers, in which the child does not participate, should be reserved only for severely disabled individuals who lack head, arm, trunk, and leg strength or are profoundly retarded. Parents who must lift a severely impaired child should become familiar with the principles of good body mechanics to protect themselves before they begin a lifting regimen. A Hoyer lift may be purchased to use with the obese child who cannot aid in the transfer.

This chapter examines only a few of the problems involved in bringing up an intellectually normal child with a birth defect. Many children with such defects require orthopedic procedures that are not described here in detail. Their relative lack of mobility may cause these children to be more subject to infection. Many children with birth defects also have diminished intelligence, learning disabilities, behavioral problems, or epilepsy. Each of these complications in essence is treated as if it existed independently and is discussed in another chapter of this book.

Few disorders tax a parent more than those of the child with a birth defect because so much can be done for these children, and the quality of their lives is so dependent on the excellence of their treatment. However, because of his critical needs, the raising of a child with a birth defect who can participate in the world around him is a source of enormous gratification for parents and involved professionals.

REFERENCES

Allum, N. 1975. Spina bifida: The treatment and care of spina bifida children. Allen and Unwin, London, p. 168.

Anderson, E. M., and B. Spain. 1977. The child with spina bifida. Methuen Co., London, p. 352.

McLaurin, R. L., ed. 1977. Myelomeningocele. Grune and Stratton, New York, p. 850.

Stark, G. D.,1977. Spina bifida: Problems and management. Blackwell, Oxford, p. 192.

Swinyard, C. A., ed. 1966. Comprehensive care of the child with spina bifida manifesta. Institute of Rehabilitation Medicine, University Medical Center, New York, p. 147.

Swinyard, C. A. The child with spina bifida. Ibid. p. 29.

Woodburn, M. 1975. Social implications of spina bifida. 2nd ed. NFER Publishing Co. First published by Scottish Spina Bifida Assn., p. 275.

16

MENTAL RETARDATION

The term mental retardation is applied to a heterogeneous group of people. We have adopted the American Association on Mental Deficiency's definition: "Mental retardation refers to significantly subaverage intellectual functioning existing concurrently with deficits in adaptive behavior, and manifested during the developmental period." As defined by the Association, mental retardation includes subnormality in two dimensions, intelligence and social adaptation. Mental retardation can be associated with physical handicaps or seizure disorders, but it need not be. The child with mental retardation can be classified according to the severity of the disorder. In the past, terms such as idiot, imbecile, and moron were used to define mentally retarded children. Apart from being insulting, these terms had very little meaning with regard to the child's ability to learn and progress. With the advent of intelligence testing, children are now classified according to their performance for age; an I.Q. below 80 is considered below normal. Borderline intelligence is defined as an I.Q. between 70 and 79; educably retarded is defined as I.Q. between 60 and 69; and trainably retarded as performance between 50 and 59. Despite the endpoint of this scale, many children with I.Q.s below 50 can be trained in self-help skills and in the elements of appropriate behavior. Furthermore, it is clear that intelligence as measured by tests does not represent a fixed capacity but rather a developing intellectual and social competence that can change with age. The President's Committee on Mental Retardation suggests that it is appropriate to require both an I.Q. of less than 70 and a substantial failure in adaptive behavior to categorize an individual as retarded.

THE FAMILY PHYSICIAN

As we have indicated, the younger the child, the more difficult it is to predict that he will be retarded unless there is a well-defined underlying disease. Retardation may be easily identified in a child of one or two years, yet the degree of impairment may still be difficult to assess accurately. Consultation with a specialist who is familiar with these problems often produces more detailed information about prognosis and treatment, although the cause may never be determined. In general, a consultation with a neurologist will be most helpful to a child at the time the diagnosis is first suspected because the underlying condition may be treatable. A single consultation with a specialist in child neurology is usually sufficient to establish whether more extensive clinical tests need to be done. This may require a brief period of hospitalization. If a child's retardation should prove to be the result of a metabolic abnormality that can be modified either by diet or medication, it may be necessary that the care of the patient be transferred to a physician experienced in this area. These disorders are unusual, and treatment of the metabolic problem must be very precise. Only a small number of physicians are familiar enough with these disorders to undertake their long-term care.

Several years of observation on the part of the family and their physician and the help of other professionals such as a psychologist may be needed before the degree of retardation can be determined. Over the years, the more closely a child's development matches that of the normal child, the more comfortable the family physician should be about predicting future outcome. A physician's initial hesitancy about giving a prognosis should be looked upon as a hopeful sign rather than as inability or unwillingness to inform parents.

Any physician who undertakes the care of a retarded child should be acutely aware of associated problems which, if left untreated, will hamper development. He should alert the parents to be watchful for signs of impaired hearing or vision. The incidence of hearing loss and visual defects is higher in retarded children, and this relative deprivation of sensation may seriously hamper the child's already limited potential. Retarded children also are more apt to develop epileptic seizures than normal children. Families should be aware of this possibility; they should understand that it is important to inform their physician about sudden changes in their child's behavior. Parents should be advised about "child-proofing" their home, that is, making absolutely sure that medicines, poisons, or objects that may cause injury are kept out of the reach of children. Retarded children, of course, are less easily taught to avoid hazards. At the same time, parents should not restrict the child unnecessarily, since retarded children need more, not fewer, opportunities to learn.

Frequently, other problems associated with raising a retarded child will be troublesome and the family will require advice from the child's physician. Irregular bowel habits, toilet training, poor eating habits, and behavioral problems are foremost among these. In many instances, these problems can be solved through the use of behavior modification techniques, which can be taught to parents by professionals skilled in this area.

As the child grows older, educational programs must be planned for him. Infant stimulation and early education programs are available in most parts of the country at present. The benefits of these programs are still controversial, but the better programs certainly help the parents care for a retarded child. Special school districts exist in most metropolitan and many rural areas. Requirements for enrollment in special education programs vary from state to state, but in most cases, a formal evaluation of the child's intellectual level is demanded before he can be admitted to a program. In most communities, the evaluation can be accomplished through the school system, but in others parents may want to assume responsibility for a private evaluation. Consultation with specialists in educational planning is indicated. Often children whose I.Q.s are not measurable or fall below the accepted trainable range can be taught some basic social skills that will make them easier to live with. Frequently, this requires temporary placement in a special training school. At any rate, the words "trainable" and "educable" should be looked upon as descriptive but *not* as absolute categories which will forever limit a child.

Often there is some fear that the child is retarded because of an inherited disease. If an exact diagnosis can be made, other members of the family can be tested for the problem and defined as either carriers or as totally normal. If the mother becomes pregnant again, it may be possible to test for the disease *in utero* and to abort the pregnancy if the family feels it is advisable. In most children, however, no specific genetic diagnosis will be established, and it will not be possible to inform other members of the family if they are carriers of a disease.

The family of a retarded child has the right to expect their physician to be able to discuss sexual relations, marriage, and vocational training with them and to advise them about preventing their child from having problems in those areas. If the retarded child marries, he would be more likely to have children who are of normal intelligence than to have children who are also retarded. However, if his children are normal, a severe family problem is often created with regard to raising them. This must be considered by the physician and appropriate advice given to a retarded individual who wants to marry. In general, individuals who are retarded and seek marriage are in the upper educable range of intelligence. They may be working, and their social adjustment may be excellent. Whether such a person can handle the responsibilities of a family life is a judgment that must be made by the individual and his family. Perhaps it is more important that the retarded child's family learn about contraceptive devices so that the child can be protected from unwanted pregnancies. Many children who are retarded are physically attractive, but because their judgment is poor they can be easily led into a potentially distressing situation. Therefore, it is imperative that parents discuss these topics with their family physician.

Frequently, the care of a retarded child requires an enormous amount of time and effort; as a result, other family members may feel neglected, and relationships within the family may become strained or even destroyed. A family physician is in a unique position to give advice in this area or to refer the family to a professional trained in family therapy for counseling. One of the most frequent reasons for the institutionalization of retarded children is the inability of the family to care for the child without seriously infringing upon the lives of other members of the family. This is particularly true of the emotional adjustment of siblings. It is of extreme importance that these matters be discussed frankly with all family members. The need to institutionalize a child on this basis should not provide a great deal of guilt. A second problem that often leads to institutionalization later in life is intractable, aggressive, and destructive behavior. Behavior of this kind may develop from an organic cause and thus possibly can be treated with drugs. Frequently, however, such behavior develops because the family's behavioral expectations for a retarded child differ from those they may have for their normal children. Parents may fail to set reasonable limits for the child and as a result he fails to learn how to function within society. This is a grave mistake that often begins early in the child's development and should be anticipated by the family physician and discussed with parents.

THE MEDICAL SPECIALIST

The specialists often involved in the evaluation and treatment of children with mental retardation are pediatric neurologists, geneticists, psychiatrists, or physicians who are interested in the endocrine and metabolic problems of children.

There are a number of reasons for referring a retarded child to a pediatric neurologist or geneticist:

1. To try to determine why the child is retarded. The family of a retarded child is appropriately concerned about the cause of the disorder, but this can be extremely difficult to determine. Problems during pregnancy and delivery are associated with a somewhat higher incidence of retardation in the child. Some of these factors have been mentioned on page 17. Since many children who are subjected to similar insults develop normally, it is difficult to establish cause and effect in any child. Even if there is a history of a mishap in pregnancy or in the perinatal period, unless it is severe and well documented, other causes that may be treatable should be considered.

Physicians are particularly interested in determining whether a child with mental retardation has an inherited disorder or an acquired disease that is progressive and can be helped by early intervention. Children whose condition appears to be deteriorating may have a metabolic or endocrine disorder. If parents bring their child to a specialist, they should certainly expect the child to be screened for inherited disorders and thyroid disease. Some of these diseases are not associated with typical physical findings and can be diagnosed only by laboratory tests. Many can be diagnosed and treated early in life with benefit to the child. If a child has periodic seizures or brief periods when he appears to be more difficult to arouse, the levels of certain chemicals in the blood (calcium, magnesium, and glucose) should be measured. It is particularly important to be sure the child does not suffer from low blood sugar (hypoglycemia), since this is a correctable problem which, if left untreated, can result in further brain damage.

Endocrine disorders are often acquired, but metabolic diseases such as disturbances of amino and organic acid metabolism are usually inherited. If such a disorder is discovered, parents should be informed that any future children have about a 25 percent chance of having the same disease. If the metabolic defect is known, it is possible in some instances to determine *in utero* if the unborn child is affected and treat the child early in life, or, if the family wishes, abort the pregnancy. It may also be possible to determine if other siblings carry the disease and could transmit it to their children, although the siblings themselves may not show clinical symptoms. Genetic counseling is extremely important in preventing the transmission of many causes of mental retardation.

Children whose neurological condition worsens gradually may also have an inherited metabolic disorder. Frequently, these disorders involve complex combinations of fats and sugars that are important to the structure of brain cells. Many of these diseases can be identified by chemical tests that use the child's blood cells or cells from the skin (fibroblasts). Occasionally, a biopsy (obtaining a small sample of tissue) is necessary to establish a diagnosis. However, there is no instance that we are aware of in which a biopsy of the brain is of benefit to the child except for certain rare infections, since inherited disorders that can only be identified by brain biopsy cannot yet be treated. If this procedure is being considered, it should always be presented to parents as a procedure to help other family members who may wish to have children in the future. If a brain biopsy is suggested to parents in order to determine why their child is retarded, they should know: (1) why the physician suspects the child has a degenerative disease; (2) whether he believes it to be infectious or hereditary; (3) if he believes it is hereditary whether he has studied other tissues to obtain a diagnosis, and if not, why not; and (4) what the purpose of the biopsy is. Can the child be helped if a diagnosis is made? If a physician answers these questions to parents' satisfaction, they can then decide whether to have the procedure done.

Many of these special laboratory studies can only be performed at major medical centers and require the participation of one or more specialists in child neurology, genetics, metabolism, or neurosurgery. Some of the tests are extremely expensive. If a pediatric neurologist suggests that they be performed, it probably means that, based on the history supplied by parents or physical findings that he has noted on serial examinations, the child has deteriorated. It is up to the neurologist to inform the parents of what has been found and why hospitalization is indicated. However, there are routine screening tests that should be done on all retarded children, i.e., blood and urine amino acids, urine organic acids, tests of thyroid function, and a fasting blood glucose. These tests can be done without hospitalizing the child.

Geneticists are also interested in identifying unusual combinations of physical findings that can occur with mental retardation. Certain syndromes are known to be inherited; therefore, if a syndrome can be identified, the family can be forewarned. Some syndromes are associated with abnormalities in the structure of the genetic material that is transmitted from parent to child (the chromosomes). If the child has distinctive physical features associated with mental retardation, a geneticist may wish to study his chromosomes to determine if their structure is normal. Blood or a small piece of skin necessary for the test can be obtained from the child when he is in the doctor's office. Most chromosomal abnormalities are the result of errors in the division of cells that eventually will form the baby and are not always inherited. However, some chromosomal abnormalities can be inherited even though neither parent has symptoms. If a child is found to have a chromosomal abnormality, the geneticist will wish to examine the parents' chromosomes. When the parent's chromosomes are looked at carefully under a microscope, one may see that the genetic material of one parent, while present in normal amounts, is not distributed properly; thus, the child is apt to get too many or too few genes from this parent. In that case, parents can be advised about the risk they face that future children will have similar problems.

In many instances, it is not possible to identify the cause of a child's retardation, but parents will still want to know what the chances are that future children will be retarded. The risk probably is slightly above that of the general population, but in fact there is little or no information on this subject at the present time. We feel that if a child's retardation cannot be attributed to a specific cause, there is probably a little better than one chance in twenty that a couple's future children will have a similar problem.

2. A visit to a child neurologist may be suggested to determine the severity of the child's symptoms irrespective of their cause. However, because patterns of development vary from child to child, it is difficult to predict the severity of retardation in very young children. Accurate predictions can be made by the time the child is between two and three years of age, but the closer the child is to being normal, the more difficult the process of prediction becomes. Unless a child's nervous system is severely damaged, a neurologist is rightfully cautious about telling parents early in their child's life what degree of retardation will be present later. Predictions that are valuable for placing a child in a treatment environment generally require the help of a psychologist (see Chapter 5).

If parents feel that their child is not developing normally and they are referred to a child neurologist, he should be able to tell them if their child's social, sensory, and motor skills are within the normal range, or if they differ significantly from other children his age. Often his judgments are tempered by the child's medical history. For example, if the child has spent a great deal of time in the hospital because of repeated infections, or was a thirty-two week premature infant, he will not be expected to have reached all of his normal developmental milestones at age six months. Time is the crucial factor in this situation. If, over the span of one to two years, differences between that child and other children his age become more apparent, then a more precise prediction can be made. On the other hand, if the differences become less prominent, it is quite possible that the child will be normal. Even if their child appears to be developing normally, many families wish to know if the child will develop a learning disability later in life because he was premature or had some other problem at birth. A learning disability cannot be diagnosed with certainty until the child enters school and begins to read, write, and use figures.

3. A specialist in child neurology should be able to determine whether a child's mental retardation is associated with physical problems such as cerebral palsy or epilepsy. If there are associated problems, he should plan a treatment program with the child's parents.

4. Many retarded children are referred to a specialist later in life because they are not accomplishing the tasks expected of them at school or because they have behavior problems that are troublesome to the teacher and to the parents. A child neurologist in cooperation with a psychologist can usually assess whether seizures or physical handicaps are preventing a retarded child from performing at the level expected for his degree of intelligence. These specialists may also be able to determine whether the child's span of concentration is so brief that it limits his ability to learn or whether he has a specific learning disability that further hampers his academic progress.

It often takes a number of visits to establish whether behavioral problems reflect an emotional disorder. These difficult judgments are made easier if there is good communication between the school, the school counselor, and psychologist or psychiatrist, and the physician. A good general rule is that the behavior problems of borderline or mildly retarded children are generally not the result of their neurologic disease. In a few children, concentration may be improved with judicious use of medications such as Ritalin or Cylert but on the whole, poor adjustment with negative behaviors, temper tantrums, and lack of interest generally results from the child's not being expected to behave properly simply because he is retarded. In more severely retarded children, behavioral disturbances are more often the result of brain damage. Some of these children benefit from treatment with medications familiar to both the pediatric neurologist and psychiatrist. However, severely retarded children also suffer from behavioral problems that reflect their upbringing. Retarded children who are potentially capable of an independent or semi-independent existence in society, but who have serious behavior problems, would benefit if their parents' used the techniques of behavior modification to teach the child acceptable ways of responding to his environment (see p. 83). Drugs cannot replace the teaching of acceptable behavior to retarded children. In those few instances where attention is improved, drugs may make behavioral training easier.

THE PSYCHOLOGIST

The psychological tests used to evaluate children suspected of being mentally retarded are the same as those used for normal children—intelligence tests were designed and standardized for this purpose. Intelligence is measured on a comparative basis. To make a diagnosis of mental retardation, a child's performance on intelligence tests is compared with the average responses of a large sample of children of the same chronological age. Clinical judgment and an interview with the child and his parents can help to determine if the child's performance is consistent with the parent's report of his activities and behavior at home. Tests of academic readiness or progress, perceptual-motor development, and social-emotional adaptation should be included in a thorough evaluation of a child thought to be mentally retarded. The focus of the evaluation is to identify any areas of strength that could be capitalized on to advance the child's progress in school.

For the purposes of educational planning and motivation, the child's performance on each test is considered in terms of his potential. For vocational planning, the child's skills must be considered in terms of the level of skills enjoyed by others he will have to compete with in the job market. Thus, an eighteen-year-old retarded youngster who, despite adequate teaching is unable to read beyond a second-grade level but enjoys working outdoors and knows a great deal about the care of plants would be best served if his vocational planning were aimed at preparing him for a job that drew on his areas of strength rather than on the skills he lacks. The purpose of the psychological assessment is to provide a complete profile of the child's skills or lack of them on measures of intelligence, academic progress, social-emotional adaptation, self-help, and vocational aptitude. With this kind of information, educational and vocational plans can be made that will enhance his prospects for employment in the future.

CASE 16.1

BT was an eleven-year-old girl who was being cared for by a child neurologist for her problem of a congenital static encephalopathy (a nonprogressive disorder of brain function apparently present from the time of birth) manifested by psychomotor retardation and seizures when first referred. BT had been identified by her doctor as language-delayed and mentally retarded at four years of age. However, her parents were reluctant to accept this diagnosis and felt that in a "good school" BT would "catch up" to other children her age. As a result, BT attended a number of different private schools, some as a "day" student and others as a full-time residential student. She had been in a residential school for the last two years. BT had been evaluated repeatedly both in the schools she attended and by private psychologists. These evaluations consistently produced test scores placing BT in the educable mentally retarded range. Stanford-Binet and WISC-R I.Q.s of 69, 61, 57, and 60 had been obtained earlier. With one or two exceptions, the professionals involved in BT's developmental assessments over the years agreed that the child's language delay reflected her global retardation rather than an isolated language deficit.

The parents were particularly concerned about BT's inability to make friends. They felt that she had never been accepted in a group and thought that this would be imperative if she was to develop a good feeling about herself. In addition, BT's parents expressed frustration and concern about their inability to cope with BT's increasingly defiant, argumentative, willful behavior. BT's neurologist referred the family to a consulting psychologist who had a particular interest in the educational and behavioral problems of mentally retarded children. She was experienced in the use of behavior modification techniques with retarded children and taught parents how to use these techniques to increase desirable behaviors and eliminate or reduce inappropriate behaviors.

Psychological evaluation

BT was a slight, pleasant, right-handed, 11-year-old child who had an awkward, stumbling gait. Mild perseveration, some drooling, impulsivity, and emotionally labile behavior were noted. BT was easily distracted from the task and often interrupted the examiner to relate some rambling, irrelevant story about something that had happened at home or elsewhere. BT liked to talk and kept up a fairly constant flow of rather meaningless chatter. Social judgment was poor and thinking quite concrete. Eye-hand coordination was poor. The following test results were obtained: BT's performance on the Peabody Picture Vocabulary Test, a measure of receptive language skills, lagged severely behind her chronological age (chronological age, 11.2 years; mental age, 6.6 years). On the WISC-R she earned a Verbal I.Q. of 60, Performance I.Q. of 61, and a Full Scale I.Q. of 60. An analysis of BT's WISC-R profile revealed functioning within the mental defective range of intelligence. All scores were well below average, with no particular strengths or weaknesses noted. These test results indicated an eleven-year-old who was functioning within the educable mentally retarded range.

Educational assessment

Wide Range Achievement Test scores: Reading (word recognition): grade placement, 1.6; percentile, 3; Spelling: grade placement, 1.2; percentile, 2; Arithmetic: grade placement, 1.0; percentile, 2.

The Woodcock Reading Mastery Test was at first-grade level.

Judging from these test results, BT was learning at a rate consistent with her measured mental age. She could write her name, and read letters of the alphabet and two words of the word recognition test. Spelling skills were limited to copying beginning letter shapes with the proper orientation. Arithmetic skills included counting and correct identification of single-digit numbers and several double-digits.

Visual-motor performance

BT held a pencil with her right hand. She demonstrated below-average performance on paper and pencil tests of eye-hand coordination such as the Bender Gestalt and the Benton Visual Retention Test.

Parent interviews

The parents complained that BT's behavior was becoming a problem. She was often irritable, had temper tantrums, and tended to be defiant and argumentative if asked to do something. They stated that their biggest problem was knowing what they could realistically expect from her. They were beginning to wonder whether her behavior was a reaction to their frustration and concern or her way of "getting back" at them for sending her away to school.

The parents were still eager to find the best educational opportunities for BT, but they had begun to question whether BT's problems would be solved if only they could find a "good school." Clearly they had become somewhat more realistic about their child's potential and more receptive to the idea that she was mentally retarded.

They did have some positive things to say about their child and described her as generally warm and affectionate. They talked about BT's future and what would happen to her when they were no longer around to take care of her. Further questioning revealed that BT really had not had many opportunities to learn to interact with peers socially because she had either been away from home at school or in special programs, or she had attended small private schools with children who did not live nearby. BT did like to be with people and wanted very much to attend a local school.

Comment:

BT was identified as educably mentally retarded when she was four years old. However, it was many years before her parents could accept this diagnosis and its ultimate implications. During that time, BT had many tests which confirmed that her I.Q. was in the educably retarded range but the family continued to search for a therapy, a program, or a school that would make BT normal. Some of these "treatments" were of questionable value. Others were good programs but inappropriate for this child. BT attended a succession of private schools with little to show in the way of achievement. She had actually made no progress in simple self-help skills other than dressing, and she had no friends. There was no indication that BT had a learning disability, although this possibility was constantly entertained by her parents.

When BT was tested at age 11, her performance was that of an educably retarded child. She had no specific areas of difficulty; there were no large differences between her verbal and performance I.Q. or in subtests on the WISC-R profile. Furthermore, her abilities as assessed by the Wide Range Achievement Test was consistent with that expected from her WISC-R.

These were not new findings. What was new was the parents' increasing concern about BT's behavior. The fact that BT was so slow to learn age-appropriate behaviors despite their efforts to rear her properly made them begin to acknowledge that BT was indeed mentally retarded and that much of her unacceptable behavior was a reaction to undue pressure. They were therefore ready to accept more realistic goals for their child and to involve her in programs that would concentrate less on academic skills and more upon control of behavior, improving social skills, and emphasizing those practical aspects of learning that she would need in everyday life, such as how to handle money and how to find her way about a city on a bus. As they became more accustomed to the idea that BT was retarded, their objectives became more realistic.

The psychologist and the parents discussed possible educational alternatives for BT. Although the parents knew that the public school system provided special schools for retarded children, they had not, in the past, considered this appropriate for their child. As a result they were unaware of the many fine programs available in their district. The psychologist talked with them about the district's special classes and recommended that the parents make contact with the school administration and arrange to visit the programs available for BT. BT was subsequently enrolled in a full-time curriculum appropriate to her level of functioning where she was considered to be making excellent progress. At an appropriate time, BT was channeled into a suitable vocational training program that equipped her to work in a sheltered workshop. By the age of eighteen, BT was reading, albeit slowly, at third-grade level. She had learned to make change and was attending training sessions at a skills center in preparation for work in a sheltered workshop. In the meantime, the parents continued their behavior management sessions with the psychologist and were pleased to find that when they applied the techniques they learned, BT's behavior improved markedly. The psychologist encouraged the parents to join the local chapter of the Association for Retarded Children. As a result of the parents' activities in this association, BT began to meet other children like herself and from similar backgrounds with whom she could interact. She soon became involved in their social activities, enjoyed being with them, and behaved appropriately in social situations.

At the age of twenty-one this young lady is being prepared to take up residence in a group home where she will be well supervised yet be relatively independent.

THE EDUCATOR

Since Public Law 94-142 has gone into effect and IEPs are now required to specify the educational needs of individual children, there is a greater tendency to group mentally retarded children in classrooms by these needs, rather than by assigning them to school placement according to arbitrary, inflexible I.Q. categories.

Parents should be aware that two separate criteria are used to describe mentally retarded children. Performance on a reputable standardized intelligence test (e.g., Wechsler Intelligence Scale for Children, Stanford-Binet Intelligence Test for Children) yields information about the individual's potential ability to learn new material, as well as his capacity to use previously learned information to cope with complex ideas and concepts. But adaptive behavior is also very important. Thus, a child who has no behavior problems, is socially aware, and has developed appropriate self-help skills might benefit more from spending some time in a regular classroom than another child with a higher I.Q. whose behavior is disruptive, who lacks necessary self-help skills, and whose adaptive behavior is poor.

The process involved in determining the correct placement for a retarded child is the same as that outlined earlier in the sections on education. First, the child must be identified as retarded. With the more severely retarded child this diagnosis is usually made quite early, either by the physician, or on the basis of parents' observations and confirmed by the doctor. However, for children who are less impaired, detection might not come until later. This usually occurs when the child starts nursery school or perhaps a bit later, when he enters kindergarten and the teachers notice that he is very slow to learn or cannot keep up with the rest of the class. Intervention should be started as soon as it is clear that a child is retarded. Associations such as the National Association for Down's Syndrome or the National Association for Retarded Citizens are excellent sources of information about appropriate preschool programs. The child's physician might also be able to direct the family to good local programs.

When a child is of school age, he is entitled to a "free and appropriate" education provided by the local school district. All of the regulations and protections included in P.L. 94-142 apply with equal force in the case of a mentally retarded child. Parents who disagree with decisions resulting from the IEP conference may use all the due process procedures included in P.L. 94-142 to initiate an impartial hearing for their mentally retarded child.

The specific curriculum and the techniques used to teach a mentally retarded child are determined solely by the child's own needs and abilities. Programs vary according to the requirements of the children within it. Children whose demonstrated intellectual potential places them in the moderately retarded range (I.Q. less than 50) and who show very poor adaptive behavior will be offered training in prosocial behavior, self-care, and self-help skills. The more intellectually able children are offered a program that includes education in the basic school skills (e.g., reading, spelling, arithmetic). While many mentally retarded youngsters do master some of these basic skills, their achievements are generally limited. A realistic goal is that each child will learn those self-help and academic skills that are within his capacity. Educational planning for those children who are less impaired should include placement in a program designed to develop vocational skills that will lead to employment and partial or total independence. Programs include practice using public transportation, marketing, cooking, proper use of household appliances, etc. Job placements are available when the student completes his education. Some graduates arrange to live in houses or apartments with a resident counselor who helps them with any difficulties they might have. Others might need more sheltered situations and closer supervision.

Because of P.L. 94-142 and its ruling that handicapped children must be evaluated prior to placement and reevaluated periodically, there is little chance that a child will arbitrarily be labeled as "retarded" and placed in a special classroom indefinitely. Parents should make sure that their child is reevaluated periodically and that his educational program is updated to fit his changing needs.

THE PHYSICAL AND OCCUPATIONAL THERAPIST

The mentally retarded child has a right to physical and occupational therapy in accordance with his special needs. Referral should be made by a physician. Training programs attempt to provide the mentally retarded person with life situations that simulate those he will encounter in society. Physical and occupational therapists emphasize skills of daily living, motor coordination, and the use of purposeful, goal-directed activities to complement the objectives of such programs.

All daily living tasks, whether dressing, eating, using the toilet, riding a bike, playing with a friend, participating in school, or working at a job, require certain basic skills. These include: gross motor skills, the ability to move within the environment; fine motor skills, the ability to use the eye and hand together; perceptual motor skills, the ability to interpret and respond appropriately to information coming in through our senses; and personal-social skills, the ability to communicate with others meaningfully. These are the skills that physical and occupational therapists should work to refine or develop in the retarded child. Therapists also participate actively with other members of the team in behavior management programs by helping to carry out a consistent program throughout the day.

To establish an individual's program, the physical and occupational therapist would first evaluate the child's level of function, estimate the child's needs based on this assessment, and then plan and implement a treatment program to help him upgrade his basic skills. For example, the therapist working with the retarded infant or young child might aim at helping the child develop motor skills such as sitting, rolling, and reaching while trying to minimize nonproductive motor habits such as repeated arching of the back or repetitive rocking. The therapists would promote early play and exploration and encourage early cooperation in dressing and feeding as well as appropriate social interaction. Parent education and involvement in such activities is a crucial ingredient for consistent progress. In addition to educating parents about their child's program, the therapist may help them recognize what they can realistically expect of their child and how they can help foster appropriate motor, functional, and social behaviors. Parents of young retarded children may need support in accepting the possibility their child's progress may be slow and gains small.

With the school-age child, the therapists attempt to teach how to perform specific activities of daily living. These skills are not limited to dressing, toilet training, feeding, and bathing. They include any activity appropriate to the child's level of mental function. For example, writing may be an important task to a child in school, but he may be extremely clumsy in his control of fine motor movements or in eye-hand coordination. The therapist can help him learn to hold and use a pen or pencil with greater ease. This will not automatically increase his written vocabulary or his recognition of letters or words, but if he does have some facility with language, it will allow him to use that ability with fewer frustrations and perhaps will increase his interest in becoming more proficient.

Most therapy techniques will build sequentially from simple steps toward more complex skills and will be based on the normal sequence of development. For example, a child would learn to put on his shirt before learning to button; he would learn to grasp a spoon before learning to color or write; he would learn to understand "no" verbally or behaviorally before being expected to respond to simple or complex commands. Sometimes skills can be taught more readily by the use of adapted techniques such as individually designed methods of dressing or feeding, by use of special equipment, or by altering the environment to meet the special needs of the child. For example, a child with a poor grasp can drink more easily using a cup adapted with a cuff fitting on the palm.

In vocational training centers or sheltered workshops, the occupational therapist may be involved with the assessment of potential occupational skills such as the ability to follow directions, to work eight hours a day, and to follow routine procedures. Depending on the facility's purpose and the abilities of the retarded person, the occupational therapist may be responsible for training the individual in specific skills and in the general routines needed for a given type of work. An environment that simulates the retarded individual's intended area of work with graduated levels of difficulty should be included in the program.

Within the residential setting, the therapist should be involved in organizing and implementing daily routines for the institutionalized person that are comparable to those of the nonretarded population, such as dressing, eating, leisure, work, and bedtime schedules. "Normal" habits are promoted, such as neat attire, maintenance of self-care, correct walking patterns, and acceptable behaviors. Partial or complete integration into the community is a goal of many residential settings and their programs aim at preparing the retarded individual for some degree of productivity and participation in the community.

Physical and occupational therapists may also be responsible for providing appropriate leisure or recreational schedules for the retarded, such as outdoor programs, crafts, cooking, or physical education activities. Excessive leisure and nonproductive time-filling play are no longer viewed as the only activities retarded children or adults are capable of achieving or enjoying. "Forget the mind, train or entertain the hand" is an outmoded attitude that demeans and keeps the retarded person from participating in and achieving a sense of worth, happiness, and value in our culture.

These are some of the many avenues of help that parents have a right to expect to find in physical and occupational therapy programs for mentally retarded children.

Retarded children frequently require services that are not supplied through traditional channels: the physician's or psychologist's office, the school or an institute specializing in physical, occupational and speech therapy. Additional services at the community level were recognized as central to the adjustment of the retarded child or adult over twenty years ago by the National Association for Retarded Citizens (NARC). Fortunately, in some parts of the country many of the programs they proposed have been enacted. They suggested:

1. Community diagnostic-treatment clinics with parent counseling.

2. Home counselors to assist parents in home management.

3. Special nursery classes and classes with normal children where possible.

4. Special education and vocational training for educable children and adults, as well as social training for those who were felt to be trainable.

5. Sheltered workshops.

6. Community day care centers for the more severely retarded.

7. Vocational rehabilitation with a goal toward placing some citizens in regular employment.

Facilities and opportunities vary in a given community. Those available are generally known to regional social workers employed at the local general hospital or a community mental health center, at school, or through the local branch of NARC. The family of a retarded child should make full use of *all* available community services to ensure that their child will lead the fullest life possible, given his intellectual capabilities. Quasi-public groups operating at the community level have had their greatest successes obtaining assistance to modify the effects of mental retardation.

REFERENCES

Baumgartner, B. B. 1975. Helping the trainable mentally retarded child. Teachers College Press, Columbia University, New York, p. 97.

Carter, C. H., ed. 1978. Medical aspects of mental retardation, 2d ed. C.C. Thomas, Springfield, Ill., p. 919.

Classification of mental retardation. 1972. Amer. J. Psychiat. 128:11, Suppl. p. 45.

Copeland, M., L. Ford, and N. Solon. 1976. Occupational therapy for mentally retarded children. University Park Press, Baltimore, p. 226.

Freeman, R. D. 1978. The use of drugs to modify behavior in retarded persons. A practical guide for parents and advocates. National Institute of Mental Retardation of Canada, Downsville, p. 25.

Furness, S. R. 1974. Education of retarded children: A review for physicians. Amer. J. Dis. Child. 127:237–242.

Gray, M. Z., ed. 1977. Mental retardation: Past and present. President's Committee on Mental Retardation, U.S. Government Printing Office #040-000-00385-1, p. 276.

Johnson, V. M., and R. A. Werner. 1975. A step-by-step learning guide for retarded infants and children. Syracuse University Press, Syracuse, N.Y., p. 195.

Koch, R., and K. J. Koch. 1975. Understanding the mentally retarded child: A new approach. Random House, New York, p. 301.

Matheny, A. P., Jr., and J. Vernick. 1969. Parents of the mentally retarded child: Emotionally overwhelmed or informationally deprived? J. Pediat., 74:953–959.

Mittler, P., ed. 1977. Research to practice in mental retardation, Vol. I: Care and intervention. University Park Press, Baltimore, p. 472.

Popovich, D. 1977. A prescriptive behavioral checklist for the severely and profoundly retarded. University Park Press, Baltimore, p. 431.

Rutter, M., and L. Hersov, eds. 1977. Child psychiatry: Modern approaches. Blackwell, Oxford, p. 1024.

Sarason, S. B., and J. Doris. 1979. Educational handicap, public policy and social history: A broadened perspective on mental retardation. The Free Press, New York, p. 460.

U. S. facilities and programs for children with severe mental illnesses. 1977. U.S. Department of Health, Education and Welfare, DHEW Publication No. (ADM) 77-47, p. 504.

Wilkin, D. 1979. Caring for the mentally handicapped child, Croom Helm, London, p. 223.

17

NEUROMUSCULAR DISEASE

A motor deficit is the inability to use one or more parts of the body appropriately; it does not necessarily involve disease of the central nervous system. However, damage to the nerves or muscles may also result in severe problems. Permanent damage to nerves can result from a birth injury; temporary damage can result from inflammation or nutritional problems, many of which can be corrected. Inherited nerve disorders may also be expressed as progressive motor deficits. Damage to the muscles may be permanent and progressive, as in muscular dystrophy. A specific diagnosis is needed if the physician is to provide a family with meaningful information about the child's ability to function in the future.

THE FAMILY PHYSICIAN

Disorders of the nerves and muscles are often extremely difficult to diagnose. In some instances, these diseases can be successfully treated with medication if a specific diagnosis is made. Many of these disorders are inherited, and the pattern of inheritance varies with the specific diagnosis. As a result, diseases of the peripheral nerves or muscles are usually diagnosed and treated by specialists in neuromuscular disease. Even if such a specialist is not immediately available in the area in which the family lives, it is usually worth the effort to travel to a major medical center to seek consultation if this type of problem is suspected.

THE MEDICAL SPECIALIST

Diagnosis and treatment of neuromuscular disorders can be very complex and are best handled by specialists in neurology, orthopedics, and genetics. Some neuromuscular disorders appear quite severe in infancy and result in a very floppy, weak, loose-jointed, hypotonic child whose later development is, nevertheless, relatively normal except for some residual clumsiness. Other disorders are progressive and fatal. Many, but not all, are inherited. The diagnosis of most neuromuscular disorders can be made more specifically later in childhood than in infancy. The child must be carefully examined and undergo some laboratory tests, including a muscle and nerve biopsy and electrical studies of muscle and nerve tissue. This may require a brief period of hospitalization.

If the specialist diagnoses the child's problem as a disorder of nerve or muscle, he should be prepared to answer the following questions:

1. What is the disease and is it hereditary? A neurological examination combined with electrical testing of nerve and muscle function and a muscle and nerve biopsy can usually distinguish between disease that begins in the spinal cord, in the nerve itself, or in the muscle. Frequently, it is more difficult to make a diagnosis in a very young, floppy infant. Several visits may be required for the physician to be sure of the exact nature of the problem. Nevertheless, even when the child is only a few months old, the distinction between a disease that is primarily muscular and one that involves the nerves or the spinal cord can generally be made.

An exact diagnosis is important because many diseases of the muscles and some diseases that involve the nerves or spinal cord are hereditary. A muscular dystrophy may be suspected by the doctor from information provided by the parents and from physical examination of the child. Dystrophic muscle leaks chemicals into the bloodstream, and identification of certain enzymes that are in part derived from muscle, such as CPK (creatine phosphokinase), can be helpful in confirming the diagnosis. A definite diagnosis may than be made by a muscle biopsy. A number of disorders of the muscles that are manifest early in childhood can be diagnosed only by the examination of samples of specially stained muscle tissue under the microscope or, in some cases, by use of the electron microscope, a highly complex instrument used to magnify tissue many thousandfold.

The most common progressive muscle disease, Duchenne's muscular dystrophy, is inherited through the mother, who does not exhibit physical evidence of the disease. A mother who has one son with Duchenne's dystrophy has a 50 percent chance of transmitting the disorder to her male children. Half of the daughters will be carriers (they can transmit the disease to their male children). Occasionally, disorders that resemble Duchenne's dystrophy or other types of muscular dystrophy are inherited as autosomal recessives, which means that both parents contribute to the expression of the disorder, and that any child in that family will have a 25 percent chance of having the disease. Carriers of Duchenne's dystrophy (the mother or sisters of the boy who has the disease) can sometimes be identified by having their serum CPK levels measured. The test is not infallible and should be done at least three times. However, if one or more levels are moderately high there is a good chance the female is a carrier.

Since there are many kinds of muscular dystrophy, precise information about the specific type a child has is essential. This helps predict the rate of worsening and the pattern of inheritance, and to determine whether there is a method to identify others in the family who are capable of transmitting the disease. This is important genetic information that a specialist should be able to help parents obtain.

2. What is the treatment of the disease in its acute and chronic stages? Some diseases of the nerves and muscles are initially very acute but once the initial stages are over they then become chronic and require long-term care. If a child has a disease such as myasthenia gravis, which affects the ability of the nerve to make the muscle contract, or if he has an inflammation of the nerves or muscles, he may become extremely weak over a matter of hours or days and not only have trouble using his limbs, but difficulty swallowing and breathing as well. The parents' introduction to a specialist under these circumstances may be in the emergency room of a hospital or in a hospital intensive care unit. Diseases of the nerves or muscles that start this acutely usually remit or can be treated. Therefore, the physician's major concern during the period of time when the child is acutely ill is to see that he is kept breathing and gets enough oxygen. This may be ensured by placing a tube into his trachea and using a respirator. It is usually best to do this procedure while the child can still breathe "electively." The parents' permission will be needed for this surgery, and the physician should explain to them all the hazards involved as well as the potential benefits to the child from protecting his airway. He should also emphasize the hope that the use of a respirator and a tracheal tube will be necessary only temporarily.

During this period of acute illness, it is unlikely that a physician trying to save the child's life will have a great deal of time to explain all the long-term problems associated with the child's neuromuscular disease. However, when the child has stabilized, these problems should be discussed with the parents in detail, including the possibility of acute recurrences. Nerve and muscle disorders that appear this rapidly usually are not hereditary.

The parents will want to know if their child's neuromuscular disorder can be treated. Diseases that attack the cells of the spinal cord are generally not treatable, although if they are due to viral infections they may be self-limiting and some recovery may be possible, as in the case of poliomyelitis (polio). Others are relentlessly progressive and result ultimately in death because of decreased movement, impaired breathing, and an increased chance of infection.

A number of nerve diseases can be treated. For example, some inflammations of the nerves themselves or of blood vessels supplying the nerves may respond to antiinflammatory medicines such as steroids. Nerve disease caused by compression of the nerve by bone or by a mass such as tumor may be relieved by surgery. Other nerve diseases can be treated if the cause can be found. For example, those types of nerve disease associated with kidney failure or diabetes may improve if the underlying disease is controlled. Similarly, if the nerve disorder is due to poisoning, such as by arsenic or some other chemical, gradual recovery may occur if the cause is removed. Recovery may be only partial and can take months or years. If no cause is found for the nerve disease, spontaneous recovery may still take place. It should not be expected that a physician will use many drugs to combat the disorder. He will probably explain that time is the patient's best ally. Occasionally, children with undiagnosed nerve disease are given large doses of vitamins in the hope that this will result in some improvement. Actually, except in those rare instances in which the nerve disease is due to a deficiency of a specific vitamin, it has not been established that large doses of vitamins combat nerve diseases.

Many forms of muscle disease can be treated. Inflammatory muscle disease can be treated with steroids or other drugs that suppress the child's immune response. Again, the side effects of these medicines are sufficiently severe that the specialist must see the child frequently. He should also warn the child and his family about complications of the medicine, such as infections, diabetes, and stomach ulcers. Some muscle diseases are due to errors in body metabolism, and it is possible to correct or at least reduce their effects. Children may develop chronic muscle disease because they have disturbances in their hormones or blood salts. Muscle disease can be seen in children with too much or too little thyroid hormone or with excessive adrenal hormones. Abnormalities of blood salts such as potassium, calcium, and magnesium can result in either transient or chronic muscle weakness. If the basic disorder is corrected, muscle strength will usually return.

Unfortunately, some diseases of the nerves and many of the common diseases of the muscles cannot be cured, nor can their progression be halted. However, these diseases can be treated by physicians with the help of closely associated professionals. The object of treatment in these disorders is to keep the patient moving and engaging in normal activities for as long as possible. As with cerebral palsy, it is important to prevent joint and skeletal deformities. Therefore, neurologists work in cooperation with physical and occupational therapists and orthopedists. Bracing, appropriate use of back supports, and when necessary a wheelchair, may be indicated to permit the patient to move about, continue to learn, and have as normal a social life as possible. With proper aids and training, the child may maintain some form of mobility for many years. In addition, breathing exercises and the use of appropriate cushions (see p. 239) may minimize skin infections associated with poor mobility. Any specialist caring for a chronically ill child who has neuromuscular disease must be attentive to these problems.

Myasthenia gravis is a disease in which the nerve is unable to excite the muscle appropriately. Since the muscles cannot contract strongly for a sustained period, the child becomes weak. Myasthenia gravis is a particularly difficult problem for children and their families to understand. Most physicians think that myasthenia gravis is an autoimmune disease. Autoimmune diseases are disorders in which the body produces substances that interfere with the functioning of its own cells. In the case of myasthenia gravis, it appears that a protein may be produced that interferes with the ability of the muscle cell to receive information from the nerve. This information is transmitted by release of a chemical, acetylcholine, from the nerve ending and its attachment to a receptor on the muscle. Normally, sufficient amounts of this chemical will cause the muscle cell to contract. The abnormal protein interferes with attachment of acetylcholine to the muscle and thus reduces the strength of muscle contraction.

While it is not the purpose of this book to discuss the theoretical basis of chronic diseases of the nervous system, it is essential that we do so when talking about myasthenia gravis. The treatment of this disorder depends directly on understanding how the disease is produced. For many years the treatment of myasthenia gravis has involved giving patients drugs that prevent the breakdown of acetylcholine. This means that more of the chemical will be available at any one time to compete successfully with a protein or any other substance that may be blocking it from its receptors on the surface of the muscle. Therefore, the muscle will contract more effectively. Unfortunately, too much acetylcholine at the site of the muscle has the same effect as too little, although for different reasons. If the receptor site on the muscle is bombarded with too much acetylcholine, it can no longer transmit information from new nerve impulses. Since the effect of these drugs is to make more acetylcholine available at the junction of the nerve and muscle, too much of such a drug has the same effect as too little—the child becomes very weak. This is a dangerous situation. Since weakness is a symptom of the disease, parents who do not understand the effect of the medication may decide to increase the dose without bringing the child to see the doctor. Parents of a child who has myasthenia gravis should understand that if the child becomes weak, he must be examined by a physician who is familiar with this disease in order to determine whether he is receiving too much or too little of the drug. Such decisions cannot be made by the family, by physicians inexperienced in management of the disease, or even by physicians who are experienced in the management of the disease but who have not examined the child. If a child with myasthenia gravis suddenly becomes weak, he should be brought to the hospital or to see his specialist quickly.

There are other methods of treating myasthenia gravis that also depend on understanding the cause of the disorder. Since it is thought that this is an immune disorder, drugs that suppress the immune reaction, such as steroids, can sometimes result in a remission of the disease. Removal of a gland in the chest, called the thymus, may also induce remission of the disease. This organ produces cells that regulate the manufacture of the protein that interferes with the attachment of acetylcholine to the muscle receptor. If the manufacture of this protein can be decreased, it may be easier to manage the disease. It is understandably quite difficult to explain to a family whose child is having problems walking up a flight of stairs that a chest operation to remove a gland may improve the strength in his legs. However, this is frequently the case. Such patients must be selected carefully and operated upon in major medical centers where there are physicians who are familiar with the special techniques necessary to care for children with myasthenia gravis before and after the chest surgery.

3. Is any new research being done that will help the child? One of the prime responsibilities of specialists who care for neurologically impaired children is to be familiar with current research and new ap-

proaches to treatment. There is probably no area where more research is being done at the present time than in the neuromuscular diseases. This has been due in great part to the fact that these diseases strike children and to the support stimulated by the Jerry Lewis Muscular Dystrophy Centers.

THE PHYSICAL AND OCCUPATIONAL THERAPIST

Most children with neuromuscular disease lose muscle strength suddenly or gradually. In many, the disorder is progressive. Others have decreased strength and tone in infancy which can improve later in childhood. It is the task of the therapist to strengthen useful muscles that may help the child move about or do his schoolwork; to prevent contractures irrespective of the potential for recovery; to discourage abnormal postures that can lead to back or hip deformities; and to keep the child mobile and independent in self-care for as long as possible by the use of braces, parapodium, crutches, wheelchair, transfer devices, and equipment to assist in arm and hand control that may allow a weak child to feed himself or even to help dress himself.

As the methods employed are not significantly different from those described in other sections of this book, they will not be discussed in detail here. Suffice it to say that the basic philosophy of caring for a child with a progressive neuromuscular disease is to encourage mobility and self-sufficiency for as long as possible. When the child has an acute disorder from which recovery is possible, the therapist's role depends upon the severity of the illness. If there is almost total paralysis, the initial goals may be positioning and passive stretching to prevent skin ulcers and contractures. As strength returns, an exercise program is begun making use of those muscles that appear to be recovering function rapidly. Later, if it is clear only partial recovery will occur, aids to assist in walking and selfcare are introduced.

The psychometric evaluation and education of children with motor deficits is discussed in Chapter 14, Cerebral Palsy. Unfortunately, many neuromuscular diseases are progressive and consequently place further restrictions on the child's activities as his condition worsens. Nevertheless, learning is not only an essential skill, but a real pleasure for many of these children and provides an outlet for their curiosity and energy. Every effort should be made to continue their participation in educational activities and to stimulate their minds despite the progressive nature of their illness.

REFERENCES

Brooke, M. H. 1977. A clinician's view of neuromuscular diseases. Williams and Wilkins Co., Baltimore, p. 225.

Dubowitz, V. 1968. Developing and diseased muscle: A histochemical study. Spastics International Medical Publications, W. Heineman, London, p. 106.

Franks, H. 1979. Will to live. Routledge and Kegan, Boston, p. 147.

Rowland, L. P., ed. 1977. Pathogenesis of human muscular dystrophies. Excerpta Medica, Amsterdam, p. 896.

Walton, J., ed. 1977. Disorders of voluntary muscle, 3d ed. Churchill Livingstone, Edinburgh, p. 1149.

18

LEARNING DISABILITIES AND LANGUAGE DISORDERS

A specific learning disability is an inability to learn to perform a well-defined academic task normally when that skill is within the capability of a great majority of individuals with comparable mental and physical abilities. There are various learning disabilities related to speech, reading, writing, the use of numbers, and other learned tasks. A learning disability should be described as precisely as possible rather than just given a name. It is also extremely important to distinguish specific disorders of learning from mental retardation or emotional disorders, which might also interfere with a child's performance at school. Such a distinction is implicit in the definition of a learning disability, as the affected child is, in other respects, intellectually normal.

In this book we have tried to differentiate language disorders from specific learning disabilities by limiting the former term to disorders that result in delayed understanding of language or delayed development of speech. The definition of a learning disability that we have chosen differs slightly from that accepted by the federal government and defined in Public Law 94-142, which deals with the education of the handicapped. That Act states: "A specific learning disability means a disorder in one or more of the basic psychological processes involved in understanding or using language, spoken or written, which may manifest itself in an imperfect ability to listen, think, speak, read, write, spell or to do mathematical calculations. The term includes such conditions as perceptual handicaps, brain injury, minimal brain dysfunction, dyslexia and developmental aphasia. The term does not include children who have learning problems which are primarily the result of visual, hearing or motor handicaps, of mental retardation, of emotional disturbance, or of environmental, cultural, or economic disadvantage."

This is the definition that most school districts use to classify children with learning problems. It is at once too vague and too specific. We do not know the cause of most learning disabilities. Thus, eliminating persons whose problems are arbitrarily declared environmental or cultural or due to economic deprivation is a hardship to the poor, in particular. Furthermore, since learning disabilities are closely associated with behavioral problems, it may be unfair to separate them. Limiting learning disabilities to those functions that disturb the use of some form of language (including mathematics) eliminates many children who can't visualize spatial relations. However, since visuo-spatial disability usually indirectly affects language and can be classed under the vague terms "perceptual handicaps, brain injury" or "minimal brain dysfunction," these children are generally eligible for benefits but at the discretion of each school district. Many children said to suffer from "perceptual handicaps" and "minimal brain dysfunction" do not have a specific disability in learning a defined school skill, although their behaviors may make it difficult for them to learn at all. It is the coupling of strictly defined disabilities in learning with these vague terms about brain function (when we do not know how the brain functions in learning-disabled children) that has resulted in such enormous variations in the diagnosis and treatment of school problems from one school district to another.

The Association for Children with Learning Disabilities has provided a list of symptoms they feel can be found in all children but appear in groups or "clusters" in children with learning disabilities. Some of these symptoms are quite specific and signal a specific learning disability to a professional examining a child. Others are so vague as to cover most children, especially those with behavioral difficulties, regardless of cause. As noted by the Association for Children with Learning Disabilities, many of these symptoms occur at one time or other during normal development; only if they persist to the detriment of the child's school performance do they assume a larger significance. Unfortunately, professionals are now being asked to see many three-year-olds who "can't follow multiple directions" or are said to be "poorly coordinated." Children who are just learning to write are also referred because they reverse some letters (a normal variant at the start of first grade).

While it is worthwhile to identify children with a specific learning disability as early as possible, over-concern about children with one or two of these "symptoms" at an early age may be alarming to the parent and detrimental to the child. Professionals need to use common sense in applying some of these terms to the individual child.

THE FAMILY PHYSICIAN

The family physician may be the first professional parents turn to for help after their child's teacher has informed them that he is not doing well in school. He should attempt to confirm the suspicion of a learning disability by evaluating the child's ability to read, write, and use numbers. It is best to use a sample of the child's classroom materials to screen for such disorders. The physician should be especially interested in testing the child's hearing and vision. However, the clear delineation of a learning disability requires the aid of several professionals, particularly a psychologist and a neurologist trained in this area. It is extremely time-consuming and often technically difficult to define exactly the extent of a learning disability; the final evaluation is best left in the hands of specialists.

The general physician is in the best position to direct the family to appropriate specialists. Too often families involve their child in time-consuming and nonproductive forms of treatment, e.g. eye exercises, programs to improve balance, etc. The family doctor can help them avoid these programs. On the other hand, the long-term care of children with learning disabilities involves problems that may be handled very capably by the family physician. It is extremely hard for parents to understand why a child who appears to be normal in every way is having such difficulty in one or more aspects of his schoolwork. Frequently, parents think of the child as uninterested, lazy, negative, or even antagonistic, and it becomes the job of the family practitioner to act as the child's advocate and to support the programs outlined by specialists in learning disabilities.

Unfortunately, if a learning disability is unrecognized and the child, the family, and the school are not prepared for the problem, the child's problem may come to light as a behavioral disorder. Such children often are negative, aggressive, and destructive because they are frustrated in their efforts to learn. It is important for the doctor to discuss the child's problems with him and decide whether he needs further counseling by a professional. A learning disability is never a reason or an excuse for inappropriate behavior. The family physician should make this clear to parents and direct them to programs that will help them deal with their child's behavioral problems. The physician and the child's teachers or counselor at school should keep in frequent contact in order that they may cooperate in planning future programs and prevent deterioration in the child's behavior.

The child who fails to develop language at the time the parents expect him to is also often brought to a physician. Since language is one of the most variable of all developmental functions, often not appearing until well after two years of age in normal children, a physician may well wait until the child is about twenty-four to thirty months old before expressing concern about his language development. Generally, parents are the first to be worried because they are most familiar with their child's language development. A physician makes several important distinctions based almost entirely on the parents' account of their child's problem, and then tries to confirm them by examination and tests. These distinctions include: (1) whether the child is retarded and his language development is proportional to his overall developmental delay; (2) a judgment about the child's hearing—if comprehension is normal it would suggest that the child hears; and (3) an evaluation of muscle function around the throat and mouth and elsewhere throughout the body to identify neuromuscular diseases that might be preventing normal speech. Most children who are not speaking by age 2½ should certainly have an audiometric examination and a psychological evaluation. It is most important that deafness be ruled out as the source of a language disorder, since it can best be treated at an early age. If a child is young and not very cooperative, electrical studies may be necessary to determine whether the child's hearing is impaired. Such tests may still be indicated if results of the psychological evaluation reveal that the the child has some degree of mental retardation. Deafness may be the result of middle ear disease, inner ear disease, disease of the nerves, or disturbance in selected areas of brain. These can usually be distinguished by further sophisticated testing. None of these tests are within the province of the family physician, but it is his responsibility to be alert to the possibility of hearing loss in a 2½ year old child who has not developed speech. The family physician should arrange to have the child's hearing and intellectual development evaluated.

THE MEDICAL SPECIALIST

The medical specialist most frequently involved in the evaluation of language disorders is the pediatric neurologist. Language disorders generally fall into two groups: speech that is difficult to understand because of poor articulation, and delayed speech. The pediatric neurologist should be able to tell the family whether the child's speech disorder is associated with a concurrent neurological problem. If it is clear that a child's speech problem is primarily due to poor articulation, he will probably benefit from speech therapy.

Children are also referred to a child neurologist because they have not begun to speak at what is deemed an "appropriate age." The most important information the neurologist can obtain is the parents' description of their child's disability. Often neurologists see children because they are not saying words at 15 or 16 months of age. This is within the normal variation of language development. If the child's physical examination is normal, and he appears to hear sounds, usually little else will be done at the time of the initial visit, but the parents should expect the child to be seen by the doctor at intervals of three to six months. If the child's hearing is questionable, he may have to be evaluated by special electrical tests that measure hearing, i.e., electric response audiometry. If a child has had repeated ear infections, he may be referred to an ear, nose, and throat specialist to determine whether he has fluid in his middle ear that can interfere with hearing or whether scarring from prior infection is reducing the transmission of sounds.

If a child is not speaking at 2½ years of age, a more extensive evaluation should be undertaken. The history obtained from the parents is still crucial. The doctor will be interested in the child's ability to use the muscles of the mouth and throat for other purposes, such as making noises, swallowing, or sucking. He will examine those muscles very carefully. He will be interested in whether the child understands language, i.e., obeys simple commands, or if he turns to or responds to soft or loud sounds. He will want to know if the child's motor skills and social behavior are developing normally. If a child is not speaking at this age a formal evaluation of hearing is usually routine. The aid of a psychologist is often sought to evaluate nonverbal as well as verbal skills to determine whether there is a significant discrepancy between speech development and other abilities. Often a formal language evaluation by a trained speech therapist is indicated.

If the child's development is normal and testing fails to reveal a cause for the problem, the neurologist may recommend that parents enroll their child in a nursery class or preschool program several half-days each week. Many normal children who have limited speech at 2½ to 3 years of age will be stimulated to speak in the presence of other children. If there is a difficulty in hearing or language processing, treatment should be started promptly under the direction of skilled speech therapists. If the child's motor skills are also delayed, it is highly likely that the "delay" in speech reflects a more generalized retardation of intellectual development rather than an isolated problem. However, the child should still be followed closely since many of these children do require speech therapy later in life.

Pediatric neurologists see many children with learning disabilities. While they are often taken to their family physician first, the initial impetus for such a referral to a neurologist may come from the school, which is concerned either about the child's failure to learn or about his behavior in class. We have noted previously that a child with a specific learning disability is by definition a child who, despite normal or above-normal intellect, finds it extremely difficult to learn one or more specific school skills. Occasionally the term learning disability is used incorrectly to describe an attractive child who is doing poorly in school but is friendly and talkative and, when evaluated by a psychologist, is found to be of borderline intelligence or mildly retarded.

If a child is referred to a pediatric neurologist for an examination to confirm a diagnosis of "learning disability," the results of the evaluation should explain (1) whether the disability is part of a more general neurologic or psychiatric disorder; (2) which school skills the learning disability encompasses; and (3) what, if anything, can be done to help the child. In order to answer these questions, a child neurologist must frequently enlist the aid of a psychologist who will also evaluate the child (see p. 292).

Many children are referred to a specialist because they are failing in all subjects with the exception of gym, music, or art. Extensive examination by the specialist typically reveals that the problem lies in one of three areas. Most frequently, it is found that the child is performing at a level consistent with his intellectual capacity which, unfortunately, may be well below that of his classmates. As a result he is unable to compete with them in school. These children are often from families who send their children to highly competitive schools whose students are usually very bright. The curriculum of these schools is often enriched and geared to the abilities of bright children. The average or dull-normal child has considerable difficulty keeping up with his classmates, and this is reflected in his grades. Such a child may do much better in a less demanding school situation or if he is put in a small classroom where he can receive individual attention and be allowed to move forward at his own pace. If this type of situation exists, the advice of an educational counselor regarding placement is essential.

Families frequently resist this idea and insist that their child continue in the same school. They reason that the child will be stimulated by classmates who are brighter than they are. Actually, the opposite is usually the case. The child is trying hard and performing to the best of his abilities, yet he never achieves any success. After years of repeated failure despite considerable effort on his part, the child loses interest in learning anything, looks upon any form of education with extreme distaste, and frequently becomes a behavior problem at school. Children generally do much better at school and enjoy the educational process much more when their efforts are rewarded by some degree of success. A neurologist, psychologist, or educator who suggests that the pace of learning be tailored to the child's abilities is acting as the child's advocate. Their advice is based on a knowledge of the unfortunate results that have occurred when children have been forced to remain in a situation that is hopelessly beyond their abilities.

Recommendations for placement must obviously take into consideration the educational facilities that are available to the child. For example, if the child functions at the lower level of the dull-normal range (e.g., I.Q. 80), and there are only two types of classes available in the school system the child attends—those for normal children and those for retarded children—neither placement is ideal. Given these two choices, the child may do best if he remains in the regular class but is allowed to work at a slower pace. In this way he could maintain and further develop his social skills through his interactions with normal children. It would be very important for the psychologist, educational consultant, and medical specialist to make sure that the family understands and accepts the reasons for the placement. Placing a dull-normal child in a classroom with retarded children is undesirable if he is offered a curriculum that makes no demands on his abilities and gives him no opportunity to use the perfectly appropriate social skills he possesses and will need to maintain in order to cope successfully in society. The law now requires that specific educational plans be made to meet the needs of children with this type of problem (see page 101). The child's parents will certainly need the continued guidance of an educational counselor. This service is generally provided by the school system, but if not, the parents should ask the teacher or their child's physician to direct them to the right resource for help.

Often a disturbed child is the product of a disturbed family, and if the child's educational problems are to be satisfactorily resolved, family counseling by a psychologist, social worker, or psychiatrist will be needed. Sometimes the child is found to be suffering from depression or some other serious psychiatric illness that will require psychiatric treatment.

A child's productivity in class may be well below the expectations of his teachers and his family if his concentration is poor. Children may have a limited attention span without being hyperactive or emotionally disturbed. For example, a child's attention span may be reduced by medication. As we have noted, many anticonvulsants reduce the attention span of children although they control their seizures. Antihistamines used for the treatment of allergies may make the child drowsy during school hours, and obviously this will interfere with learning. Unrecognized seizures may interrupt a child's ability to concentrate. A specialist in child neurology will be aware of these possibilities and will either eliminate them or focus on them as possible causes of the child's difficulties and devise a remedy. Children vary widely in their ability to attend for long periods of time. Children who have short attention spans but are otherwise normal will usually improve in their ability to attend as they grow older. Some respond temporarily to stimulant medications such as Ritalin or Dexedrine in low doses. However, they often do better in highly structured settings where there are minimal distractions, and appropriate supervision and teaching methods are available to suit the child's needs.

If a child has a specific learning disability such as an inability to read, write, or use numbers, a neurologist and psychologist working together can usually assure the family that the problem is a specific one. They can also define the extent and severity of the disability and make suggestions about methods by which the child can be taught and tested that will not penalize him because of his disability. However, these suggestions, as well as the development of special programs, must be implemented through the school system and require that the neurologist, psychologist, and educational therapist work in conjunction with one another.

The ultimate goal of the specialists who are caring for a child with a severe learning disability is twofold: first, to be sure that the child continues to enjoy school, wants to learn new information, and makes appropriate use of the knowledge that he does have; second, to make certain that he is proficient enough in the area of his disability, be it reading, mathematics, or writing, to meet the demands of society. For example, a child who has serious difficulty with arithmetic nevertheless must learn to do simple calculations and make change. It probably does not matter terribly much if he knows algebra or geometry. Similarly, a child who has difficulty with reading or writing must be able to read and understand simple directions and fill out application forms that may be required for him to get a driver's license or get and keep a job. Therefore, a learning disability cannot be ignored; special instruction is essential. Often, as a child grows older and the school system becomes more flexible, bright children manage to work around their disabilities and participate in the educational process with an increasing degree of success.

THE PSYCHOLOGIST

Traditional standardized tests are the basis of the battery of psychometric tests used to assess children who have learning problems in school. The purpose of the evaluation is to provide detailed information about the child's intellectual potential, academic progress, visual-perceptual-motor skills, language development, and social and emotional adaptation. The results of the tests are analyzed to define the extent of the child's deficits as well as any particular strengths. The child's developmental progress (language, social-emotional, motor, etc.) and his achievement skills in specific areas (reading, spelling, arithmetic, etc.), are then compared with the level expected relative to the child's general intellectual potential. Very often the school system can provide an excellent psychoeducational assessment of a child at no cost to parents if they are concerned about his school progress. The evaluations are generally conducted by the school psychologist and the educational diagnostician. The goal of the assessment is to gather the information necessary to permit the teacher to design and implement a specific course of study that will serve the needs of the child.

CASE 18.1

HP, a nine-year-old boy, was brought to his family physician because of repeated complaints over the past two years from the child's teachers about the boy's poor school performance. The physician examined the child and, aside from some clumsiness, found no grossly apparent physical abnormalities. Because the chief complaint was school problems, the physician had the boy read aloud a passage from a third-grade reader and found that HP was unable to read the opening sentence. He was also unable to spell third-grade words and his writing was slow and awkward with misspellings and letter reversals, e.g., b for d. The physician referred the family to a consulting psychologist for assessment. The psychologist found HP to be friendly and cooperative throughout the evaluation. He was attentive and seemed to be making a good effort to complete each of the assigned tasks. The psychologist's assessment included a general intelligence test, tests of visual-motor organization, a receptive vocabulary test, in addition to tests of basic school skills (reading, spelling, arithmetic) and a measure of self-esteem and social adjustment. Test results indicated that HP functioned in the low-average range intellectually (WISC-R Full Scale I.Q. = 84). There was no significant difference between the Verbal (I.Q. = 82) and Performance (I.Q. = 84) sections of the WISC-R. However, the Verbal subtest scores ranged from a low of 3 (test age 6 years, 6 months) on the Information subtest to a high of 13 (test age 14 years) on the Digit Span subtest. The Performance (nonverbal) section of the WISC-R yielded scores ranging from 9 (test age 8 years, 6 months) for Picture Arrangement to a low of 6 (test age 6 years, 2 months) for Object Assembly. Tests of basic tool skills (reading, spelling, and arithmetic) showed a marked deficit in reading and spelling, but age-appropriate arithmetic performance. Reading and spelling were at a beginning to mid-first grade level. Word attack skills, oral fluency, and reading comprehension were all well below HP's third-grade placement and lower than one would expect, even from a child of dull-normal intellect. HP's pencil and paper skills were also very weak. He had difficulty copying geometric designs and his scores on visual-motor tests placed him at the 6 year, 6 month level. Judging from the boy's responses to a self-report questionnaire, his social and emotional adaptation was adequate. The boy perceived himself as having difficulties in school. However, he stated that he had friends, was liked by his classmates and was generally a happy boy. He considered himself trustworthy and felt he was an important member of his family. He did admit that he "hated school" and that he probably was not meeting his family's expectations as far as academic progress was concerned.

Information from the school indicated that HP was not a behavior problem. Teachers thought that the boy was trying, but that he was simply unable to keep up with his classmates. A group intelligence test given at the beginning of the third grade as a part of the school's regular testing program, placed the boy in the mildly mentally retarded range. However, HP's teachers noted that his arithmetic scores were average and that he did much better in arithmetic than in other areas. Moreover, they felt that he did not appear to be mentally retarded. He seemed to be able to function effectively in practical, everyday situations with his peers and was also said to be attentive in class and motivated to learn.

Interview with the parents

The interview with HP's parents revealed that HP had been tested when he was approximately 3½ years old, prior to enrollment in a nursery school class. At that time, he obtained a Stanford-Binet I.Q. of 90 and no mention was made of any specific disabilities. The boy was accepted into the nursery school where he was judged to be functioning at a level equal to his peers.

HP's parents had no complaints about the boy's behavior. They were, however, confused by his inability to perform at the same level as his siblings, who were both "A" students.

HP was diagnosed by the psychologist as a child with specific learning disabilities in reading, spelling, and writing. Her recommendation was that HP receive instruction in reading, spelling, and writing in a class for learning-disabled children taught by a Learning Disabilities Special Education teacher, with the reservation that HP be permitted to take part in nonacademic activities with his regular classmates.

Comment:

The problems that occurred in the evaluation, diagnosis, and eventual placement of HP are representative of many children with learning disabilities. HP was tested at age 3½, prior to enrollment in nursery school, at which time his I.Q. on the Stanford-Binet was 90, which is in the lower part of the average range. No specific disability was noted. However, psychometric tests given to children at the preschool level do not accurately forecast the development of a learning disability. Moreover, when HP was tested in the third grade at school, a group intelligence test was used. Group tests are designed to be administered by teachers to an entire class at the same time and are usually part of a school district's routine evaluation program. The test used for HP's class required reading, which placed HP at a serious disadvantage, and his low scores reflected his learning disability rather than his global intellect.

HP had been doing poorly at school for at least two years. However, he did not create problems in class, had friends, and appeared to be well adjusted. It was probably because his teachers thought that he was trying, but just unable to keep up because of his general intelligence, that referral to other professionals was delayed. This is not unusual. Referrals may be delayed for many years because a child with a learning disability who is well behaved is felt to be either retarded or of borderline intellect. As it happened, HP had a low-normal intellect, but specific learning disabilities significantly influenced his capacity to learn. Specific learning disabilities can be present in children whose intelligence varies markedly from the educably retarded to the exceptionally precocious, very superior range.

HP was fortunate that his family physician was interested in learning as a process, and thus did not limit himself to a standard neurological examination. The standard examination revealed only that HP was somewhat clumsy; however, when the boy was asked to read, make a sentence, and spell some words, it very quickly became apparent to the physician that the boy had considerable difficulty in these areas. Any physician who evaluates a child who is brought to him because of difficulties in school performance should actually have the child do some age-appropriate reading, spelling, and arithmetic tasks as part of his examination. In this case, the physician referred HP and his family to a consulting psychologist for a more extensive evaluation. The standard tests used at this age, the WISC-R, indicated a Full-Scale I.Q. of 84, with a Verbal and Performance I.Q. that were not significantly different, i.e., 82 for the former and 84 for the latter. However, there was a wide range in subtest scores. HP's poor performance on the information subtest was, in part, a reflection of HP's inability to read. Children with significant reading disorders may have difficulty with visual-motor organization, and in HP's case this appeared to be reflected in his low score on puzzle assembly tests and tasks requiring the use of paper and pencil. The degree of test score scatter found in HP's WISC-R profile suggests that there was an underlying problem apart from the patient's global intelligence. This was further borne out by testing his school skills, which again showed a deficit in reading, spelling, and writing, but appropriate arithmetic skills for his age.

It might be considered sufficient to establish that HP was a child of low-normal intellect with a specific learning disability, but it is not. It is extremely important in dealing with all school problems, and particularly with learning disabilities, to evaluate the child's social and emotional adaptation. Fortunately, HP had not developed any serious behavior problems. Nevertheless, he already perceived that he was different from his classmates and had more problems learning than they did. He was beginning to dislike school, even though he was not yet creating a disturbance in class. Many children with unrecognized learning disabilities are confronted by repeated failures even though they try to learn. Eventually repeated failures erode their interest in schoolwork and some children become very hostile to the school situation. All learning, even those aspects of education which are well within their grasp, becomes burdensome to them.

Having defined HP's situation, the question then arose as to where to place this child in order for him to learn most effectively, have some success, and still continue to have contact with his friends. Learning-disabled children are best taught in small groups by teachers especially trained for that task. Thus, it was quite appropriate to recommend that HP be transferred to a classroom for learning-disabled children for instruction in reading, spelling, and writing. On the other hand, by allowing him to return to his regular class for nonacademic activities, he would be able to continue his association with his friends and further develop his social skills. Many families fear that if their child is taken out of a regular classroom for special instruction he will be taunted by other children upon his return. This is rarely the case if the child is emotionally well adjusted, sociable, and has friends in his peer group. The fact that learning disabilities classrooms are in regular schools and are attended by many students at different times during the day also helps to avoid stigmatizing the child.

It was found that HP was what is termed a "visual reader." He read much better when he memorized the visual appearance of the word and did not try and translate it into sounds by dissecting each syllable phonetically. His training, therefore, consisted of the visual recognition of a basic group of words needed for everyday reading. Recently, HP has developed a strong interest in and familiarity with automobiles, and books about cars, racing, and racing drivers have been incorporated into his reading program with great success.

THE EDUCATOR

Learning disabilities have received much publicity in recent years and have been surrounded with confusion. On the positive side, many children who are truly learning disabled have been discovered and helped. On the other hand, because of confusion about what constitutes a learning disability (even among professionals in the field), many children have been incorrectly labeled "learning disabled," removed from regular classrooms, and placed in a learning disabilities classroom.

For the purposes of this section, we must use the definition included in the Rules and Regulations of P.L. 94–142, Section 121a.5, to define "specific learning disability," despite the fact that some of the terms used in this definition are vague (e.g., "imperfect ability to listen, think . . . , perceptual handicaps and minimal brain dysfunction"). That definition is given on page 284.

Learning disability classes may not be restricted to children with learning problems. They may in fact include children with serious behavior problems. These children frequently disrupt the classroom and interfere with learning, further handicapping the so-called "L.D." child.

One characteristic of a specific learning disability is that, despite a deficit in one or more areas, the child will have identifiable strengths in others. Parents cannot be certain if their child is actually learning disabled unless he has had a complete psychological assessment by a professional.

Unlike other handicaps discussed previously, many learning-disabled children are not identified until they enter first grade and have trouble learning to read, write, or count. Indeed, in our opinion, children should not be burdened with a label that suggests they have a problem learning before they enter school. "Learning disabilities" is a big business, and there is often considerable external pressure on parents to involve their child in a program that claims to correct such problems. Parents should be very cautious about enrolling their child in a program of this kind unless they have consulted informed professionals not associated with the program about the child's need for corrective procedures and have investigated the merits of the program thoroughly.

Once a child is enrolled in public school and is suspected of having a learning disability, he must be evaluated at no cost to parents. The next step is the IEP conference, where the placement most beneficial to the child's needs will be determined. Throughout his school career, the child must have periodic reevaluations to monitor his progress and to alert teachers to needed changes in his program. As discussed in other sections, if parents are concerned about the program or placement chosen for their child, they have a right, as mandated by P.L. 94–142, to a hearing.

School programs for learning-disabled children vary according to the nature of the disability. Educating these children can be a frustrating and difficult task for teachers and parents. The age and personality of the child, as well as the type of disability involved, plays a large role in the selection of the most effective teaching method. Often the solution lies in the use of alternative teaching methods. Thus, if a child has a serious problem learning to read and as a result is unable to keep up with his class in other areas that demand reading fluency, an alternative approach for giving him the necessary information should be used (e.g., records, filmstrips, or taped material). If a child has such poor eye-hand coordination that he finds it almost impossible to print or write, he could be taught to use a typewriter. Although using alternative teaching methods to compensate for weakness is frequently a good place to start, even the child with a severe learning disability must learn some basic survival skills. A child must know how to read signs, to fill out a job application, and how to count his change if he is to live independently. Therefore, every effort should be made to teach him the basic skills while he is in school.

There are many classes available for learning-disabled children, and often several children with different problems are in the same class with one teacher who is trained as a learning disabilities specialist. These teachers try to provide individual guidance for each child, since the differing nature of the children's learning problems prevents them from being taught as a group. Each child should be provided with opportunities for social interaction through group work on class projects in order to promote the development of a healthy self-concept.

With the advent of mainstreaming, it is now more common for learning-disabled children to be included in a regular classroom than was once the case. Teachers in such a situation consult with specialists to develop techniques to meet the needs of these special students. In addition, the children may spend anywhere from an hour to half of each school day working in a resource room with a learning disabilities specialist.

It is important for parents of a child with a learning disability to be sensitive to their child's limitations. At the same time, the child should not be allowed to use his deficit as an excuse for not trying. Parents should be quick to recognize and praise genuine effort and to help their child experience success whenever possible. If a disability goes unrecognized or untreated for many years or if the school and the family focus all their attention on the child's problem, ignoring his assets, serious behavioral problems can result. In both instances, school and formal learning become a frustrating experience to a child, and he may respond with negative rebellious behavior. Once such a behavior pattern is established it is very difficult to correct, even in a highly structured classroom. If the child is an adolescent, it may be better to enroll him in a work-study program aimed at developing vocational skills rather than trying to force him into an academic program. Early detection of learning problems, close cooperation between parents and educators, and the means to help learning-disabled children have increased in recent years. Social promotion is not entirely a thing of the past, but it is now usually carried out with a knowledge of the child's weaknesses and appropriate use of special help in those areas.

THE SPEECH THERAPIST

Educational planning for a hearing-impaired child depends heavily on the level of the child's language development. This in turn is influenced by the degree and nature of the hearing impairment. Before any educational decisions can be made for a hearing-impaired child, the child must have a thorough evaluation by a professional audiologist. The child's physician should refer the family to a competent audiologist; alternatively, the family could contact the closest state school for the deaf for names of audiologists in their area. The audiologist's report will provide important information about the kind and degree of hearing loss the child has. This information will be used to develop appropriate educational plans.

Hearing-impaired children who have some degree of residual sensory capacity can generally profit from the use of a hearing aid. In order for an audiologist to fit the child with an aid that maximizes his residual hearing, it may be necessary to try several types of aids over a period of time. Parents should not expect that their child's first hearing aid will fit his needs exactly.

Parents should also be aware of the controversy in deaf education about the best method of teaching these children to communicate. The oral method stresses surrounding the child with a purely oral environment, teaching oral communication through lip reading, speech, and the maximum use of residual hearing through amplified sound. Another method advocates teaching certain children to communicate through hand gestures called "signing" or "finger-spelling." A third viewpoint supports a total communication approach, and uses both the oral method and signing. Parents should investigate all three points of view and talk with an unbiased but informed professional before agreeing to placement in a program that adheres rigidly to only one method. The method used should be determined on the basis of the child's needs rather than on narrow professional interest. Children may start out being taught totally by the oral method, but if they fail to learn at the expected rate they can then be switched to the total approach. However, it is important to remember that all deaf children are not alike, and instead of insisting on one approach rather than another the aim should be to obtain the education program that best meets the child's needs.

A hearing-impaired child should be enrolled in a deaf education program as early as possible. The Central Institute for the Deaf (C.I.D.) in St. Louis, Missouri, for example, offers a parent-infant program that is designed for hearing-impaired children under four years of age and their parents. This program consists of weekly, hour-long, individual sessions that take place in a homelike setting. The purpose of these classes is to teach parents how to provide the kind of environment that will best stimulate their child's early learning. Other family members are frequently involved. Participants are shown how to encourage the child to talk and how to respond to the speech of others. Since the Institute's focus is on listening skills, parents are taught ways to teach their child to attend to the face and mouth of the person speaking. In addition to the parent-infant classes, the child after the age of two will attend C.I.D.'s nursery-school class for short sessions. This class emphasizes preparation of the child for preschool through the development of both social and behavioral skills. Meanwhile, parents continue to receive training in intellectual and language stimulation techniques to be used at home. Although such elaborate and extensive early training programs are not available in all states, most states do provide some deaf-infant education or stimulation programs. Parents should locate and investigate the early intervention programs available in their area as soon as their child is diagnosed as hearing impaired so that therapy can be started promptly.

Regardless of the method of communication taught, education for hearing-impaired children differs from that of other children in that during the early years, communication and language skills are emphasized. During the later years of elementary, junior high, and high school, the emphasis is on high-content subjects (i.e., science, math, geography). Consequently it is not unusual for hearing-impaired children to lag behind children with normal hearing in these subjects until the last years of high school.

Whether a child with impaired hearing should attend a regular classroom is an important issue. The determining factors are the child's language skills and what aids the classroom is prepared to offer. Frequently, local school districts hire individuals to provide deaf children and their teachers with various supportive services. A hearing clinician, for example, might meet with a regular classroom teacher twice a week to help her develop the skills necessary to work with a deaf child in her classroom. In addition, the hearing-impaired child might go to a resource room daily for extra help, which could include visual presentation of information that was presented orally to others in the class. Overhead projectors and amplification systems can be used. Volunteer groups frequently provide interpreters or note-takers for such students. There are a variety of ways that a hearing-impaired student who is otherwise normal can be helped to succeed in a regular classroom. If as a result of an IEP conference a hearing-impaired child is placed in a regular class, but parents feel that the child's needs are not being met, they can use the procedures set forth in P.L. 94–142 to make their concerns heard. At this time they will have the opportunity to discuss what services they feel their child needs and is not receiving.

Whether a hearing-impaired adolescent goes to a college, university, or vocational school usually depends on his intellectual ability, the degree of academic proficiency he has reached, and his interests. If the adolescent chooses to go on to college, he should write to the Deans of Admissions of the colleges he is interested in, explain his situation, and ask for information about special television or other equipment or special privileges (such as taping lectures or substituting papers for class participation) he might require before making application for admission. The family should contact the local Division of Vocational Rehabilitation so that a trained counselor can explain to parents and student alike exactly the kinds of services that are available, and whether funds are allocated by their state to finance higher education. Gallaudet College in Washington, D.C., distributes an excellent book about opportunities for deaf individuals.

Defining the characteristics of a learning disability frequently requires the skills of many specialists: the physician, the psychologist, and the educator. Treating a learning disability is the function of the educator and the child's parents, with the physician and the psychologist in an advisory position unless behavioral problems have become serious. Parents must be aware of their child's needs and investigate class size, pupil-to-teacher ratio, curriculum, opportunities for special educational programs, the behavior of classmates, and the discipline exercised in class to promote learning and constructive behavior. Parents must ultimately decide whether their child's educational placement meets what they feel to be his needs, and parents must defend the child's right to an appropriate education. They may, however, need the advice and even the testimony of a physician and/or a psychologist. It is also the parents of learning-disabled children who must fight to change inappropriate laws and who must make an effort to promote and fund research into learning disorders. There is probably no area in neurology where so little is known and in which so little research is being done to correct the deficits.

REFERENCES

Coopersmith, S. 1967. The antecedents of self-esteem. W.H. Freeman, San Francisco, p. 283.

Douglas, J. W. B. 1966. The home and the school. Macgibbon and Kee, London, p. 296.

Kosloff, M. A. 1979. A program for families of children with learning and behavior problems. Wiley, New York, p. 450.

Levine, M., R. Brooks, and J. Shonkoff. 1980. A pediatric approach to learning disorders. Wiley, New York, p. 339.

Patterson, C. R. 1971. Families. Research Press Co., Champaign, Ill., p. 143.

Stewart, M. A., and S. Wendkos. 1973. Raising a hyperactive child. Harper & Row, New York, p. 299.

Williams, F. J. 1970. Children with specific learning difficulties. Pergamon Press, Oxford, England, p. 211.

Zifferblatt, S. 1970. Improving study and homework behaviors. Research Press Co., Champaign, Ill., p. 96.

19

HYPERACTIVITY

"Hyperactivity" is a troublesome term that is much overused. In strict usage, it refers to a cluster of behavioral characteristics that include short attention span, impulsivity, poor judgment, restlessness, a low frustration tolerance, and overactivity in all situations. Many children have only one or two of these symptoms. They cannot really be considered hyperactive, although any one of the symptoms may impair their ability to function effectively in school, at home, or with peers. Again, it is more important to describe the maladaptive behavior than it is to affix a label to the child.

"Minimal brain damage" is a term that is often used to describe children who are overly active, who may be somewhat clumsy, and who have mild school problems, although they are not truly retarded. In our opinion, this term describes nothing but a constellation of vague symptoms and signs that often vary from examination to examination. Nothing in the term leads to a prognosis or a plan for treatment. Furthermore, it suggests that the brain is actually damaged, a claim that has never been proved. Destruction of brain tissue implies that the child's problems will be very difficult to remedy. Even those who use the term "minimal brain damage" would agree that this is not the case and that it presents a falsely bleak picture to a family. These problems are better described in detail than they are summarized by such an unfortunate phrase.

"Developmental delay" is another "wastebasket" term. It implies that a child who is slow when first examined will be able to catch up with his peers sometime in the future. The rate of development varies among individuals, and some children may in fact appear to be slow early in life yet go on to develop normally. However, there is no reliable method of predicting which child will ultimately fit into this category. The term is often incorrectly used to reassure a family about their child's problems when, in fact, they should be concerned because early intervention, if provided at a critical time, might improve development.

THE FAMILY PHYSICIAN

Most children who are brought to physicians because they are said to be hyperactive have one or more features of that symptom complex, but do not suffer from hyperactivity *per se*. Children who are labeled hyperactive often are only overactive. The degree of activity displayed by children varies enormously; some children require relatively little sleep and have great physical energy while others require more sleep to maintain the same activity level. Not all children need to sleep for ten to twelve hours even though parents might wish that their child do so. Overactive children can frequently concentrate for long periods of time on activities that interest them, and even more frequently are able to pick up information though they appear to be inattentive. Other children who are described as hyperactive may have a limited attention span, but this is not associated with impulsivity, the need for little sleep, aggressiveness, or lack of fear. Still others have behavioral problems characterized by frequent temper tantrums, or they are aggressive, negative, or fearless. At times these traits are seen at home while the child behaves normally with friends or at school. There is often no reason why a medical specialist is needed to catalogue the symptoms that may trouble a supposedly hyperactive child. This can be done by the family physician if he takes a detailed history. Furthermore, he is best able to decide whether a child's behavior really represents hyperactivity or whether it is a manifestation of turmoil within the family, difficulties at school, or, as is sometimes the case, the result of inappropriate expectations on the part of the child's parents or inconsistent discipline. He should be able to indicate alternative causes for the child's behavior and be aware of which behavioral problems require concern and immediate attention. Certainly aggressive, destructive behavior is foremost among these.

CASE 19.1

DB, a three-year-old boy, was brought to his family physician because his parents said they could "no longer stand him." He slept only four to five hours a night and frequently awakened at five o'clock in the morning. He was "into everything"; he had burned himself several times on the stove, but seemed to learn little from his experience and repeatedly returned to play near the stove. He had temper tantrums and found it difficult to play with other children. At times if they would not allow him to play with their toys, he would pick up one of his own toys and smack them with it. Recently, he lifted his four-month-old sister out of her crib, put her in his wagon, and ran recklessly down the street to show her to a neighbor who lived a block away. When the parents were asked why they had not complained of his behavior previously, they indicated that they had been concerned, but felt he would outgrow it. However, the incident involving his baby sister frightened them so that they decided to seek help immediately. The physician examined the child and found no abnormalities. He suggested that the child be referred to a consulting psychologist for evaluation. The psychologist found the child to be intellectually in the above-average range. A program of behavior modification was suggested, involving rewards for appropriate behavior, specific interventions for destructive behavior, and "benign neglect" for overactivity. In three months the frequency of the child's temper tantrums had been reduced and his overall behavior had improved.

Comment:

This child could be regarded as hyperactive. Neither the physician nor the psychologist found evidence that the child's behavior was related to serious emotional problems or to a disruptive family situation. Since his behavior had been a problem before the birth of his baby sister, this too could be ruled out as a cause. In interviews it became clear that the parents of DB had reinforced many of his undesirable behaviors by initially treating them as cute. Thus, attempting to modify this child's behavior by involving the family in an educational program was a sensible strategy.

The treatment of hyperactivity is controversial. For many years it has often included the use of stimulants such as amphetamines or Ritalin. There is no question that such drugs can improve the attention span of a small group of children. However, objective studies fail to substantiate the benefits of medication for most hyperactive children. Diets that are free of sugar and/or food additives have also been suggested. Once again, the most objective studies fail to support claims made for these treatments. Even if the diet seems to be effective, it often requires changing the family's entire pattern of living to center on the afflicted child and his dietary requirements. It is reasonable to suppose that this shift in family attitudes may be the source of the diet's beneficial effects.

Most children who have elements of the hyperactivity syndrome benefit from educational assistance, behavior modification, or both. This means that the family physician once again must act with other professionals to make certain that the child receives appropriate continuing care. Occasionally, children with elements of this syndrome will have such severe behavioral disturbances that psychiatric care will be indicated. At times institutionalization in a highly structured environment for a brief period of time may be advisable. If the problem is that severe, the family physician should participate in the decision, but he probably should seek the guidance of a neurologist or psychiatrist familiar with hyperactivity before suggesting this course to a family. Here again, one of the family physician's major functions is to alleviate guilt on the part of the family by explaining the clinical advantages of institutionalization, realizing that these active, disturbed children can stimulate great hostility among other members of the family. It must be made clear to both the child, if possible, and to the family that this course of action is being pursued by them as a form of therapy and not a form of punishment.

THE MEDICAL SPECIALIST

Children who are thought to be hyperactive are often referred to pediatric neurologists or to psychiatrists. In our experience, most children called hyperactive do not fit the strict definition of that term. Many of them have learned to manipulate their families or those around them by the use of negative behavior. These children's temper tantrums serve to structure their lives and that of the family in a way that suits them. While it is difficult to think that a three-year-old or a five-year-old manipulates his parents, this frequently happens. Hyperactivity is often a synonym for unacceptable behavior that serves as a method of gaining attention. The pediatric neurologist generally has little to offer these children or their families except to assure them that there is no underlying neurological disease and to refer them to other professionals expert in behavior modification. Similarly, children whose behavior reflects an emotional disturbance should be referred to those who can give them appropriate psychiatric care. There remains a small group of children who are truly hyperactive. Some of these children are severely retarded and must be treated with drugs designed to reduce their activity level so that they can live at home and participate in some group activities. Others are of average or above-average intelligence, but have enormous energy, cannot concentrate, are impulsive, have little understanding of what is dangerous to them, are frequently aggressive and destructive, and are always restless and require little sleep. These children display such traits whether they are at school, at home, or with neighbors or relatives. A child who is hyperactive at home but not at school or not at his grandfather's house should not be labeled hyperactive. In this case, family counseling rather than medical treatment is probably in order.

Some parents prefer having their child diagnosed as hyperactive because this suggests the child has an organic problem and therefore cannot help misbehaving. This removes the responsibility for changing the child's behavior from the family. Many families dislike hearing their child is not hyperactive or not suffering from minimal brain damage, and they search for a specialist who will finally tell them what they want to hear. They will orient the life of the family around a special diet for the child rather than focusing on his unacceptable behavior. The shift in family attitude may result in an improvement of his immediate symptoms, but there is some concern about how these children will function in adult life when they are placed in group settings that will not cater to their individual needs. No studies as yet have reevaluated individuals later in life who were treated by diet for hyperactive behavior when they were children. Behavior modification still seems to be the best technique available at this time for treating an overly active child who has poor social habits, since it offers the best prospect for helping him learn to manage his own behavior and to develop into a productive member of society. The use of drugs or diet can supplement behavioral techniques, and in a small number of children it may definitely be beneficial.

A family should be suspicious of a specialist who sees hundreds of hyperactive children, since it is doubtful that hundreds of hyperactive children live in any but the largest urban communities, and it is even less likely that they all find their way to one office. Families should also be skeptical of the specialist who offers simple solutions to aggressive, restless behavior and quickly dismisses the child as having an organic illness such as "minimal brain damage."

THE PSYCHOLOGIST

The most important factor in the psychological assessment of a hyper-active child is the skill and the experience of the examiner. Generally there is no need to modify tests or test items to accommodate these easily distracted, easily frustrated, impulsive children. However, there is a definite need for a psychologist who can supervise the child's ef-forts, set limits for acceptable performance, and provide encourage-ment and support throughout the test session. These children need many more "breaks" in the test sessions than is usually required. Testing should be divided over a three-visit span if the child is ex-tremely active and difficult to focus on the tasks at hand. If the exam-iner is skilled in behavioral approaches to hyperactive children, these techniques can often be used to advantage during the testing periods. The psychological assessment of a hyperactive child should include a detailed interview with the parent. This is needed in order to coordi-nate appropriate management at home and at school. The desired end result of the assessment process is the appropriate placement of the child in an effective and therapeutic school program.

CASE 19.2

KP (aged 7 years, 2 months) was an overactive, easily distracted, often will-ful boy referred for psychological evaluation by his family doctor. The re-ferral was prompted by the parents' growing frustration with repeated phone calls from the child's teacher about KP's disruptive, noncompliant behaviors in class. When KP arrived for psychological assessment, his teacher had sent along a letter describing the boy's behavior and performance in school. According to the teacher, KP was impulsive and careless. His work was untidy and often incomplete. She also said that KP's attention span was extremely short and that he rarely finished any assigned work. He disturbed the other children and interrupted classroom activities with his silly, clown-ing antics. His quick and almost uncontrollable reaction to frustration or discipline frequently led to his being sent to the principal's office. However, the teacher noted that KP seemed to enjoy being out of the class and, in her opinion, considered being sent to the principal a treat rather than a punish-ment. The teacher described his behavior on the playground as "wild." She said he had friends, but he sometimes became so wound up and demanding that even his friends would refuse to play with him. On the other hand, his behavior during physical education was quite good. The P.E. teacher was a young man who was firm, yet warm, with KP. The classroom teacher noted that this young man seemed to know how to handle the boy. Both the school principal and the teacher felt that KP was a very bright boy and they were genuinely seeking advice on ways to help him.

During the psychological testing, KP was compliant but was noted to be restless, easily distracted, and impulsive. He grabbed for test materials, was up and down in his seat, and often responded before the examiner had completed the instructions. He tended to answer without thinking and needed many reminders to slow down and think. He was quick to "give up" if he encountered difficulties; however, by using a clearly defined, firm approach and a reward system, the psychologist was able to maintain KP's interest and motivation. Nevertheless, he required many more "breaks" in the testing sessions than are usually necessary for a child his age. In fact, three, rather than the usual two, visits were scheduled for the evaluation to ensure optimal performance.

When tested on the Wechsler Intelligence Scale for Children (WISC-R), KP's Verbal I.Q. was 127, Performance I.Q. was 117, and Full Scale I.Q. was 125. Subtest scores clustered fairly closely together, with no areas of significant weakness. KP's fund of information was strong, his vocabulary was well developed, he was able to reason logically, and his ability to handle abstract concepts was quite advanced. The ten-point difference between the Verbal and Performance I.Q.s was not considered significant. KP's nonverbal scores, with one exception, were above average. He was quick to recognize part-whole relationships and he had no difficulty with nonverbal abstract concepts. His lowest score occurred on the coding subtest. KP was less adept at copying unfamiliar symbols presented in a nonsequential order within a time limit than he was at other nonverbal tasks, but even here his score was in the average range. Pencil dexterity was perhaps not as smooth as it should have been, but no serious deficit was noted.

Academic skills, as measured by the Jastak Wide Range Achievement Test, were as follows:

	Grade Placement	Educational Quotient (E.Q.)	Percentile
Reading (word recognition)	3.4	118	88
Spelling	3.1	114	82
Arithmetic	3.1	115	84
Gray Oral Reading Test (oral fluency)	4.6 (Instructional at fifth to sixth grade level)		
Woodcock Reading Mastery Test	4.1		

KP was an excellent reader. He was a second grader who was reading at a fourth-grade level. His reading comprehension was good. Spelling and arithmetic were equally strong. Although KP was not productive in school, he obviously had obtained needed information and mastered basic educational skills, and the results of these tests indicated that he clearly had the ability to be competitive in an academic setting.

KP's perceptual-motor skills were less advanced than other abilities. He had adequate perception, but relatively immature pencil control. In addition, KP's tendency to work in a hasty, careless manner contributed to the low scores he earned on the Bender Gestalt and the Benton Revised Visual Retention Tests. However, no gross distortions, problems with orientation, or perseverations were noted that would raise the question of a neurological deficit.

KP's social and emotional adjustment, as measured by the Piers-Harris Self-Rating Scale, indicated that KP considered himself a happy person and that he had a fairly good opinion of himself. However, there was some evidence suggesting that he was not as pleased with his school performance as his general "I'm OK" manner would indicate. Thus, he admitted that he daydreamed in class, misbehaved in the school setting, and that he was sometimes the last to be chosen for games.

In talking with KP's parents, it was found that initially they both tended to blame the school and KP's teacher for the boy's difficulties. They felt, on the one hand, that the teacher was too young and that she just let KP run over her. On the other hand, they volunteered that schools were only interested in making children into robots, and that any child who gave them difficulty was immediately identified as a troublemaker and sent to a special school. They did, however, indicate that KP had always been active, and when he was three years old they had taken him to a psychiatrist who diagnosed him as "hyperactive" and "unable to help the way he behaves." The boy had been tried on a course of stimulant medication, but there was little or no improvement in behavior and the medication was discontinued. Asked to describe the boy's behavior at home, the parents admitted that he was a terror. He was explosive, demanding, slow to do any assigned chores, and more often than not, disobedient. He often had tantrums when he could not get his way and smashed his toys. He required about five hours of sleep each night. On the other hand, he was considerate of, and got along well with, his younger sister.

Comment:

The psychologist summarized KP as a bright, easily distracted, energetic, but undisciplined boy whose intellect was in the superior range. Academic skills were well developed. KP did have a very mild problem with pencil control, but this was not felt to be a major factor in his school difficulties. Behaviorally, KP was overactive, but he did respond to structure and clear expectations for appropriate behavior.

Recommendations were outlined for the teacher and included the use of a reward system for a time to encourage more appropriate classroom behavior. For example, KP could earn free time to engage in favored activities (e.g., feeding the gerbils, taking notes to the office, etc.) if he completed his assigned work on time. The teacher was to recognize and *promptly* praise appropriate behavior (e.g., "KP, you are doing a good job of sitting still").

To counteract some of KP's distractibility and impulsivity, it was suggested that his teacher make sure she had KP's attention when giving directions, and that she try to be in close physical proximity with him and to touch him lightly to help him attend. It was also suggested that the teacher try to seat KP at a desk near her, so that she could make eye contact with him easily.

The psychologist suggested that KP be told what the rules for appropriate behavior in class were and what the consequences were for breaking those rules. If a rule was broken by KP, the consequences were to follow swiftly and consistently. Time out was recommended as a consequence, instead of sending KP to the principal's office. The psychologist stressed that time out should be for brief periods (5 to 10 minutes at most) and that the cloakroom be used for this purpose. Another suggestion was that KP's assignments should, for a time, be divided into shorter sections, and that he be given one section at a time to work on and told to bring it to the teacher as soon as it was completed, gradually increasing the length of the work sections (over a space of months). This method, since it presents smaller amounts of work to be done at one time, seems to be less overwhelming to distractible children and also gives the teacher more opportunities to reinforce and praise the child for completed work. By attending to completed work, rather than nagging a child when he is not working, the teacher is rewarding appropriate rather than inappropriate behavior.

The psychologist felt that the parents could use some guidance in learning to manage KP's behavior at home, and suggested that the parents take part in a training program designed to teach them behavior management techniques (see p. 78). She also stressed to the parents that their ambivalent attitude toward their child's behaviors at school made it even more difficult for KP's teachers to help him. It is rarely true that a child can't help his behavior. Most children can be taught to behave appropriately if their families are willing to follow a consistent approach.

The parents entered into the behavior management training program eagerly and were pleased to find that KP responded very quickly to their new approach to his behavior. A few simple rules about daily routines and clear consequences for not following the rules made it easier for the parents to be consistent. Bedtime was no longer a hassle, and mealtimes began to be a pleasant time for the whole family. The tension, bickering, and stress that previously characterized much of the family's interactions were markedly reduced. KP's teacher no longer called them every time there was an upset at school. She simply took care of the misbehavior in class where it occurred.

KP will probably always be more energetic and perhaps even more impulsive at times than other children, but with the help of his parents and his teachers, he is beginning to internalize controls for his own behavior and to accept responsibility for his acts. His energy is being channeled more effectively and he is able to use his very fine intellect constructively.

KP made good progress in the second grade and was advanced to the third grade, where he continues to do well. The parents maintained contact with the psychologist and occasionally see her for "booster" sessions. KP is not a problem-free child, but by working together his parents and his teachers feel confident that they can help KP avoid serious problems in the future.

Not all children respond as well as KP, but most will show improvement in behavior if the proper behavioral approach is used.

THE EDUCATOR

Ever since the early 1960s, it has become increasingly common to label many behaviorally troublesome children as "hyperactive." At what point should an active child be considered hyperactive? Dr. Mark A. Stewart, in his informative and comprehensive book *Raising a Hyperactive Child* (highly recommended reading for both parents and teachers of children referred to as hyperactive), describes such a child in behavioral terms as "a child who is persistently overactive, distractible, impulsive, and excitable in the eyes of his parents and teachers" (p. 3). Later, Dr. Stewart goes on to add, "It is a collection of symptoms that probably should not be thought of as a medical disorder, but rather as a variant of personality. Furthermore, the boundary between the hyperactive personality and the norm or average personality is defined by the child's social surroundings" (p. 232).

Typically, hyperactive children are identified as such at a very early age. As infants, they are unusually restless, active, and often hard to comfort. Toddlerhood can be a parental nightmare, with constant vigilance required to ensure the child's safety as well as the home furnishings and objects in the child's immediate environment. Once the child enters preschool and kindergarten, invariably the parents begin receiving complaints about their child's disruptive behavior. When it is time for academic skills to be taught, a whole new set of problems begins, since hyperactive children often experience great difficulty in developing both an adequate attention span and the ability to concentrate. Still, there are ways to help these hyperactive youngsters to learn and profit from their school experiences. Establishing good communication with the classroom teacher is important. Parents should express a willingness to keep the lines of communication open.

When initially meeting the classroom teacher, parents should try to give her an honest description of their child's strengths and weaknesses. The label "hyperactive" should be avoided, as this has different meanings for different people. Instead, specific behaviors should be described, and the teacher should be told about what the child can do and likes to do, as well as about his problem areas.

If a hyperactive child is to be successful in school, he will need an understanding teacher who is willing to individualize lessons for him when necessary. Assignments will frequently need to be broken down into smaller, more manageable units. Directions may need to be repeated, and often told to the child individually. The child may also need to be refocused to the task with a gentle reminder. Teachers of hyperactive children must be especially alert to "catching" these children when they are demonstrating appropriate behavior and to reinforce it promptly. A structured, well-organized environment will help the hyperactive child better learn to control his impulsiveness than a more loosely organized, "open" classroom.

There is no doubt that parents of hyperactive children need to take an active role to ensure their child's placement in a classroom environment that will best meet his needs. The same recourse is available to them if they are unhappy with their child's school program as is open to the parents of other handicapped children. The first action should be an attempt to discuss their concerns with the child's classroom teacher. However, if changes in the child's program are not forthcoming, it is their right, as set forth in P.L. 94–142, to initiate an informal hearing proceeding.

Hyperactivity and attention deficits can exist in children who are good students. The brighter the pupil, the less apt the teacher is to call the behavior to the attention of the school counselor or the child's parents unless he constantly disrupts normal classroom activities. If a child's negative behaviors are only present at home, an alert professional will certainly investigate the possibility that such behaviors are not hyperactivity but the child's method of exerting control. At any rate, applying a label is less important than creating situations at home and in school that modify aggressive, destructive behaviors. This is the best way to be sure that the active or hyperactive child will use his energies constructively when he becomes an adolescent and an adult.

REFERENCES

Clegg, A., and B. Megson. 1968. Children in distress. Penguin Books, Baltimore, p. 204.
Douglas, J. W. B. 1966. The home and the school. Macgibbon and Kee, London, p. 296.

Kosloff, M. A. 1979. A program for families of children with learning and behavior problems. Wiley, New York, p. 450.

Patterson, G. R. 1971. Families. Research Press Co. Champaign, Ill., p. 143.

Shea, T. 1978. Teaching children and youth with behavior disorders. C.V. Mosby Co., St. Louis, p. 343.

Smith, J. M., and D. Smith. 1964. Child management: A program for parents and teachers. Ann Arbor Publishers, Ann Arbor, Mich., p. 102.

Stewart, M. A., and S. Wendkos. 1973. Raising a hyperactive child. Harper & Row, New York, p. 299.

20

THE PARENTS

It seems reasonable to assume that the manner in which professionals and the parents of a handicapped child interact with each other could influence the parents' attitudes toward their child. Parents who come away from a visit to a professional feeling angry, disappointed, confused, or frustrated may unknowingly express these feelings toward the child. Similarly, the professional who finds a family difficult to cope with may be less thorough in exploring areas of concern with the family. Negative behaviors are provoked when parents or professionals feel the others are not discharging their responsibilities as expected.

THE OFFICE VISIT

In the preceding chapters we have discussed the skills, duties, and responsibilities parents should expect from professionals. In this concluding chapter we offer a brief discussion of what professionals expect from parents. We believe it is important for parents to be aware of this, because it is the degree to which both groups meet each other's expectations that determines the quality of their relationship, which in turn influences the care of the child.

Fortunately, most professionals no longer expect or find it desirable for the patient or his parents to follow their suggestions with blind, unquestioning obedience. Consumer groups, organizations of parents, and other "special" groups have spent a great deal of time, money, and effort alerting the general public to their rights as patients, clients, parents, students, and so on. Awe or deference is no longer automatically bestowed on anyone holding a degree. Parents should expect to take an active part in any planning that concerns their child, whether it involves medical, educational, social, or vocational decisions. However, if parents are to be partners in planning they will be expected to become informed about issues, to think objectively, to share responsibility for decisions, to ask questions about matters they do not understand, and to be willing to accept the fact that not all questions can be answered as precisely or as quickly as they would like. Parents should also know exactly why they are seeing a particular professional. Surprisingly, many families have no idea why they are sent from one specialist to another. They either never ask, are never told or, if told, they do not understand. The fault is not entirely that of the parents; it may be that of the referring physician, teacher, therapist, or agency. Nevertheless, it is ultimately the parents' responsibility to keep themselves informed. It is their obligation to ask the professional making the referral to explain why their child should be seen by a neurologist, orthopedist, occupational therapist, psychologist etc. Having agreed that the suggested referral is necessary, parents should ask for a letter of referral that they can take with them, or make sure that one is sent promptly so that the individual for whom it is intended will have it before the family's scheduled visit. The letter should include pertinent background information about the child (developmental, behavioral, educational); the nature of the child's problem; treatment or care given; what prompted the referral; and what kind of advice or care is sought. We urge parents to request copies of all records of their child's contacts with professionals so that they have the copies at hand when they are referred to a new resource. Having these records immediately available is of tremendous help, and it will in many instances avoid duplication of procedures and thus save time and money. The information should include EEG tracings, laboratory tests, reports of psychological evaluations, interviews with social workers, and any other documents that concern the child's problem.

Parents should expect to be asked and to be willing to explain what worries them most about their child and what they expect from an evaluation by a new specialist. Often their major concerns differ from those of the referring physician, teacher, or other professional.

Professionals would like parents to take the time to discuss their child's problems with each other before the initial meeting so that they can sort out the worries they both share, along with those that are of less concern to one parent than the other. If parents disagree about the need to see a specialist or about a specific therapeutic approach, the chances are that neither the family nor the professional will feel that their visit was productive. Talking things over beforehand gives parents a chance to organize their thoughts so that they and the professional will be able to make the best use of their time together.

Occasionally a visit to a professional is prompted by the parents' desire to bring a lawsuit against an individual or an institution that they perceive as responsible for their child's condition. Professionals have a right to expect parents to tell them that, if in addition to seeking help for their child, they are also considering legal action against someone or if they have a suit pending and are looking for an "expert witness" to testify for them.

Most physicians and other professionals understand and can accept a parent's need to ask questions as soon as the initial examination is concluded. However, they do expect parents to be willing to accept reasonable explanations for having to delay some answers until the results of certain tests are known. Some parents of children who have chronic, complex, and difficult neurological problems have dealt with professionals for many years, but they still find themselves beset by frustrations, anxieties, and serious problems. Such parents may use questions as a way of expressing their anger. For example, they demand immediate answers from the "new" specialist and complain that none of the professionals they saw previously told them anything. Or frustrated parents may attend the first parent-teacher conference in a new school and immediately ask the teacher why their child has failed to make progress in the past. When the new teacher can't answer this question, they may feel she is incompetent. Such behavior makes for an unrewarding and probably transient relationship. The family again feels rejected by a professional, and the cycle of anger, frustration, and despair continues, always to the detriment of the child.

It is true that professionals, particularly physicians, often expect courtesies that they are unable to return. For example, a doctor expects a family to wait patiently if he is held up by an emergency or delayed by an extra number of difficult patients, yet he expects parents to be on time for appointments. The fact is, however, that professionals rarely simply fail to keep an appointment with a family, but many parents do not keep appointments or notify the office if they are unable to come. Chronic appointment abusers are irritating and costly to the professional, and this damages the parent-professional relationship. Therefore, parents would be well advised to remember to cancel any appointment they are unable to keep.

Professionals whose services demand the cooperation of the child (e.g., psychologists, physical therapists, medical specialists, speech therapists, etc.) expect parents to use good judgment and reschedule their appointment rather than keep it when their child is sick, feverish, or nauseated and as a result is unable, unwilling, and uninterested in taking part in any scheduled activities. Obviously, the appointment should be kept if it is with the child's pediatrician or family doctor.

Medical specialists who evaluate children often have associates who will see the family first, obtain the initial background information about the child's development, his past medical history, the family history, and other pertinent information. Some may perform a general physical examination. The specialist expects the parents to understand that he and other persons in authority consider their associates competent and well qualified to perform these duties. On the other hand, it is the specialist, not an associate, who should talk with the parents and answer their questions after he has reviewed the case with his associate and examined the child himself.

CONTINUING CARE

Meeting the responsibilities expected of them as parents is an essential ingredient to a good continuing relationship with a professional. No matter what part a professional plays in the continuing care of a child, he expects parents to take part in discussions and to ask questions if they feel that progress is not satisfactory, if they are uncertain about instructions, or if they do not understand the reasons for a change in the child's program.

Compliance is a necessary part of a continuing relationship. The family and the professional should have agreed on a treatment plan for the child. Once this is done, it makes no sense for parents to fail to do their part in carrying out the plan. If the situation changes and parents feel they have good reason to alter the plan, they should discuss it with the professional and agree on modifications. Obviously, this does not mean that parents cannot use their own judgment about an exceptional event that makes it impossible for them to comply exactly as agreed. Compliance may mean that the child takes medications at the indicated times in the prescribed dosages. Compliance may mean that the family follows through with an agreed-upon therapeutic program to passively move joints or to position the child in his wheelchair or bed while at home as instructed. It may also mean keeping a written account of certain events or behaviors. For example, if the child has a seizure disorder, the neurologist may ask the family to observe the child at home, to record any seizures that occur, what happened during the seizure (e.g., eyes rolled back, limbs jerked, wet himself), how long it lasted, and what the child did after the seizure was over. For a child with a behavior problem, compliance might mean keeping a detailed account of the family's daily activities and a description of the events that preceded and followed the child's inappropriate behavior. Compliance may mean taking the child to a local laboratory for certain blood or other tests at specific times indicated by the physician. Compliance also means scheduling and showing up for return visits at periodic intervals as requested by the professional.

USE OF THE TELEPHONE

In certain situations a professional will ask a parent to make telephone reports so that the child's progress can be monitored. A doctor, for example, might want to know about a child's response to a new medication. A teacher and a parent might be working together to improve a child's school grades by keeping in touch about homework assignments. There are any number of reasons why a telephone report from a parent would be helpful, and professionals appreciate and respect parents who take the time to make the call as requested. Parents frequently have questions that they feel do not require a special office visit.

Professionals expect parents to use the telephone efficiently. Parents can do this if they:

1. Have specific questions in mind before calling (it is a good idea to write them down).

2. If a specific event led to the telephone call, the parent making the call should know the details so that he or she can answer questions quickly rather than making a professional wait while they go ask someone else what happened.

3. If the matter is not urgent, the parent should not demand to speak to the professional in person immediately. He or she should be willing to let the professional return the call when he is free. A message should be left including *the child's name*, the parent's name, and the *telephone number* where the caller can be reached. Parents should ask approximately what time they can expect the call to be returned. They should make an effort to be there to accept the call so that the necessity for repeated, useless phone calls is avoided. If possible, the telephone should be kept free during that time period so repeated busy signals don't frustrate the caller. Professionals expect parents to be willing to come to the office if their questions cannot be answered intelligently by phone.

As families become used to the idea that their child has a chronic handicap, they begin to assume greater responsibility for the child's care. Professionals then assume the role of advisors who are called on periodically to make adjustments in the child's program in cooperation with the family. This situation develops when parents have either independently or in cooperation with the appropriate professional made the effort to develop realistic short- and long-term goals for their child.

HOSPITALIZATION

When a handicapped child needs to be hospitalized, for example, for seizure control, an orthopedic procedure, etc., professionals expect parents to explain the need for hospitalization to the child, to tell him the truth about what needs to be done and why it is being done. Obviously, explanations must be within the limits of the child's ability to understand. This means that it is the family's responsibility to know when, where, and why the child is being hospitalized. The family must also understand the risks of any procedures the child might have to undergo.

Hospitalization can be very expensive. If a family needs financial aid, they should let their physician or the hospital social worker know about it rather than depriving the child or his siblings of necessities. There are many ways in which assistance can be given to a family if the appropriate professionals are alerted to the problem.

When a child is hospitalized, the physician in charge of the child's care expects parents to come to him for information rather than to question any available hospital staff member. Parents who engage in this kind of behavior are apt to get differing opinions about the child's progress. Some may be unrealistically optimistic, others may be unrealistically pessimistic, and some may be totally inaccurate. Such indiscriminate questioning of staff results in confusion, unnecessary distress, anxiety, and even anger and hostility on the part of the parents when they are given conflicting reports. Moreover, it interferes with parent-professional relationships because it can provoke conflict between the physician in charge, other hospital staff members, and parents.

Eventually, it is expected that the family, with the advice of their medical consultants, will learn to plan objectively and then will take the responsibility for making those plans work by applying them to their child with love and sensitivity.

INDEX

abnormalcy, definition of, 13–16

academic accomplishments, parents and, 114–15

accessibility, for handicapped, 174

achievement, neurological handicaps and, 17

achievement tests, educational, 65–66

activities, for epileptic child, 194–95

activities of daily living, 250–52, 272–73

adaptive development
developmental schedules and, 62
mental retardation and, 258, 270
tests for, 73–74
see also development; language development; social development

administrative procedures, in testing handicapped child, 69

agencies, private, concerned with neurological disabilities, 156–59

alcohol consumption, handicaps and, 18

American Association on Mental Deficiency, 258

American Speech and Hearing Association, 132

Ammons Full Range Picture Vocabulary Test, 210

Apgar score, as handicap predictor, 19

Association for Children with Learning Disabilities, 113, 285

bathing, 252–55, 253–54 *figs.*

Bayley Scales of Infant Development, 63 *fig.*

behavioral disturbances
learning disabilities and, 285
mental retardation and, 265–66

behavioral management, 83–84
for neurologically handicapped, 46
special education teachers and, 115–16

behavior modification
handicapped child and, 84–94
hyperactive child and, 305–6

Bender Visual Motor Gestalt Test, 72

Benton Revised Visual Retention Test, 72

Binet, Alfred, 56–58, 60

Binet-Simon Intelligence Test, 57

birth, predictors of handicaps at, 17–23

birth defects, 17, 227–56

body alignment, 242–43

bracing, 243–46, 244–45 *figs.*

brain, growth in infancy of, 7, 12

brain damage
cerebral palsy and, 201, 203, 207
insults and, 12

brain damage (*continued*)
minimal, 302, 306
neurological handicaps and, 31–32
Brazelton, Barry, 68
Bureau of Education for the Handicapped (U.S.), 101

California Mental Maturity Scale (CMMS), 211
Cattell developmental schedules, 62–63
causes, in definition of diseases, 41
Celontin (drug), 185
Central Institute for the Deaf (C.I.D.), 299
cerebral palsy, 17, 201–25
definition of, 12
evidence of, 23
sensory modalities and, 125–26
childhood, development of handicaps in, 23–25
child-parent interactions, 118, 215–16
child rearing, counseling and, 83
chromosomal abnormalities
development of handicaps and, 21
mental retardation and, 264
see also inheritance
cigarette smoking, handicaps and, 18
Civil Rights, State Regional Office of (Department of Health and Human Services), 174
clinical nurse specialists, 163–65
community resources, 34
for neurologically handicapped, 46
social workers and, 139, 150–51
computerized tomography (CT scan), 29, 207
confidentiality, 141–43
Constitution, Fourteenth Amendment to, 167
consumer organizations, 113–14, 314
continuing care
by medical specialists, 38–49

of neurologically handicapped, 34–36
by nurses, 161–65
parents and, 317–18
convulsive seizures
abnormal development and, 14–15
as handicap indicators, 17–18
maturation and, 24
mental retardation and, 258
see also epileptic seizures; febrile seizures
counseling
genetic, 42
handicapped and, 78–83
for neurologically handicapped, 45–46
of parents, 231–32
counselors
psychologists as, 79–83
social workers as, 140
CPK (creatine phosphokinase), 277–78
Crippled Children's Services, 82, 147–48
cyanosis, 19

day schools, special, 103
deaf education, 299–300
defects
birth, 17, 227–56
speech, 131–32, 287–89
vision, 13–14
deficits, motor, 276
see also cerebral palsy; neuromuscular diseases
Denver Developmental Screening Test (DDST), 7, 8–9 *fig.*, 13, 68
development
evaluation of, 47, 63 *fig.*
postnatal, 3–16, 4–6 *table*
variation in, 302
verbal, 12–13
see also adaptive development;

evaluation (*continued*)
  by school districts, 170–71
  by speech therapists, 132–33
  by teachers, 106–12
  by therapists, 123–27
exercise programs, 246–47
  see also range of motion exercises
extremities, range of motion of,
    124–26

families
  counseling needs of, 82
  diagnostic role of, 24–26
  emotional support for, 143–44
  questions of, about handicapped
    child, 40
  see also parents
Family Educational Rights and Pri-
    vacy Act of 1974, 171–72
family physicians, 27–36
  medical specialists and, 37–40
  role of, in abnormal development,
    13–16
  role of, in treating birth defects,
    227–29
  role of, in treating cerebral palsy,
    202–3
  role of, in treating epilepsy, 184–89
  role of, in treating hyperactivity,
    303–5
  role of, in treating learning dis-
    abilities, 286–87
  role of, in treating mental retarda-
    tion, 259–61
  role of, in treating neuromuscular
    disease, 25–26, 276
  therapists and, 118
febrile seizures
  in children, 15
  see also convulsive seizures; epi-
    leptic seizures
Federal Rehabilitation Act of 1973,
    173–74
feeding techniques, 235, 238 *fig.*
financial assistance

  for handicapped child, 177–80
  social workers and, 144–50
financial support, of handicapped
    child, 174–76
Ford, Gerald, 101
Fourteenth Amendment, 167
Frankenburg, William, 68
Frostig Developmental Test of Visual
    Perception, 72

Gallaudet College (Washington,
    D.C.), 300
genetic counseling, medical spe-
    cialists and, 42
geneticists, role of, in treating mental
    retardation, 263–64
German measles, *see* rubella
Global assessment, developmental
    screening and, 67–68
global retardation, 13
goals
  for child with learning disabilities,
    291–92
  of therapy, 134
  in treatment, 127–28
government agencies, handicapped
    and, 67
guardianship, of handicapped child,
    176–77

handicaps
  alcohol consumption and, 18
  at birth, 17–23, 227–56
  in childhood, 23–25
Health, Education and Welfare, De-
    partment of (U.S.), 174, 178
hearing impairments
  evaluations of, 70–71, 132
  language development and, 298
  treatment of, 14
hemiplegia, 206
Hidden Figures Test, 72
Hiskey-Nebraska Test of Learning
    Aptitude, 70

Careers For
Foreign Language Experts

Interviews by Russell Shorto

Photographs by Edward Keating and Carrie Boretz

CHOICES
The Millbrook Press
Brookfield, Connecticut

Produced in association with Agincourt Press.

Choices Editor: Megan Liberman

Photographs by Edward Keating, except: Elena Baxter (Patrick Bartholomew), Kathryn Wahlin (Carrie Boretz), Molly Brennan (Carrie Boretz), Libby Cryer (Carrie Boretz), Timothy Rogus (Steve Kagan), Maria Campbell (Carrie Boretz), J. Bruce Weinman (Carrie Boretz).

Library of Congress Cataloging-in-Publication Data

Shorto, Russell.
Careers for foreign language experts/interviews by Russell Shorto, photographs by Edward Keating and Carrie Boretz.

p.   cm. – (Choices)
Includes bibliographical references and index.

Summary: Includes interviews with people who hold such varied careers as tour consultant, literary agent, international lawyer, and foreign language textbook editor, describing what they do, how they got started, and the necessary preparation for each job.

ISBN 1-56294-159-3

1. Language and languages – Vocational guidance – Juvenile literature.
2. Linguistics – Interviews – Juvenile literature.
[1. Language and languages – Vocational guidance.   2. Linguistics.
3. Vocational guidance.   4. Occupations.]
I. Keating, Edward, ill.   II. Boretz, Carrie, ill.
III. Title.   IV. Series: Choices (Brookfield, Conn.)
P60.S54   1992   91-27661
402'.3 – dc20

Photographs copyright in the names of the photographers.

Contents

Introduction

In this book, fourteen people who work in foreign language fields talk about their careers — what their work involves, how they got started, and what they like (and dislike) about it. They tell you things you should know before beginning a foreign language career and show you how a love of foreign languages can lead to many different types of jobs.

Some of the careers in this book involve speaking a foreign language—such as interpreter, tour guide, and language teacher. Other jobs — such as literary translator, foreign language editor, and international literary scout — apply foreign language skills to writing. And some, including international business facilitator and foreign exchange director, even use language skills to build bridges between different cultures.

The fourteen careers described here are just the beginning, so don't limit your sights. At the end of this book, you'll find short descriptions of a dozen more careers you may want to explore, as well as suggestions on how to get more information. There are many business opportunities available for people with foreign language skills. If you enjoy speaking and writing foreign languages, you'll find a wide range of career choices open to you.

Joan E. Storey, M.B.A., M.S.W.
Series Career Consultant

"A second language opens doors you might never have known existed."

ELENA BAXTER

INTERNATIONAL LAWYER

Paris, France

WHAT I DO:
I'm a junior partner in a
Paris-based law firm, about
half of whose members are
French and half American.
Because the twenty-five
lawyers here have different
specialties, we have quite a
broad-based practice. We do
all sorts of corporate work,
including mergers and
acquisitions, technology
transfers, litigation, and
arbitration.

The firm's specialists in
corporate and commercial
litigation spend most of their
time in court, but I'm more
of a generalist. I typically
represent Americans and
Europeans in multi-country
business transactions. And
when I bring in a specialist
on taxation or banking, as
I do from time to time, I act
as the coordinating counsel.

Elena peruses the morning
newspaper at a favorite
sidewalk cafe in Paris.

In some ways, being a
lawyer in Paris is very much
like having a practice in New
York. The same high stan-
dards apply to the work. But
there are also real differences, the most obvious of
which concern language and
culture. When you're based in
New York, you're dealing with
the English language and
with American laws and
cultural practices. In Paris,
however, there is much less
uniformity. I work with
German, Spanish, British,
and French clients, and with
each of them, the circum-
stances are different. The way
a Spaniard approaches a
problem has nothing at all
in common with the way a
German approaches the same
problem. As a result, I'm constantly switching gears. But
that's also what makes this
work so fascinating.

A less obvious difference
is that in Europe there is
perhaps more respect for

people's private lives. It's more acceptable for lawyers here to take breaks and vacations than it is for lawyers in New York.

HOW I GOT STARTED:

At one point, I thought I was going to be a doctor. But in college I realized that I enjoyed my elective classes in political science and constitutional theory more than I enjoyed my required premedical classes.

I went to law school at the University of Virginia. Then I joined a firm that had offices all over the world. The possibility of someday working overseas attracted me. And that's eventually what happened.

I started taking French lessons when I was 8 or 9 years old, and I continued studying French through junior high and high school. I also began dreaming of other countries at that time, though it wasn't until I was 19 that I made my first trip to France. When I came back, I was determined to master the language, so I minored in French in college.

My linguistic ability was one of the things that made it possible for me to follow this career, because I have to work in two languages. I speak French with half my colleagues and clients and English with the other half, so I'm constantly switching back and forth.

Elena looks over a contract with a colleague.

Elena does some legal research for an upcoming case.

HOW I FEEL ABOUT IT:
I'm not sure I would enjoy the law as much if I were working in New York. I've been in Paris for eight years now, and that's been most of my career. Practicing international law in Paris is delightful. The quality of life is nice, and it's financially rewarding as well.

WHAT YOU SHOULD KNOW:
I cannot emphasize enough the flexibility a second language gives you in your professional life. It opens doors that you might never have known existed. It's certainly true that English is the international language, but as the world becomes more interdependent and more interrelated, you find less tolerance for people who speak only English. Twenty years ago, American lawyers came to Paris who didn't speak French. They were accepted then, but they would be much less so today.

The highest paid young lawyers are corporate lawyers in New York. In Europe, starting salaries were traditionally a lot smaller, but that's changing rapidly. Today, a young American lawyer in Europe can expect to make anywhere from $50,000 to $80,000 straight out of law school.

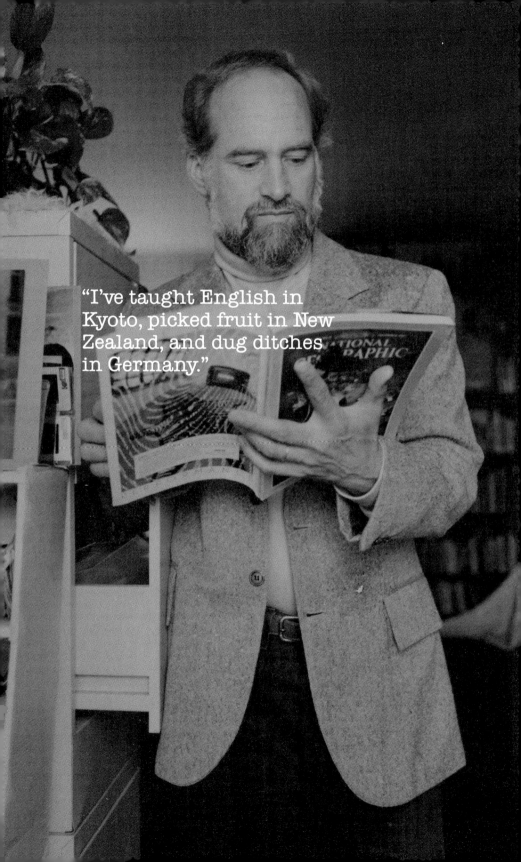

"I've taught English in Kyoto, picked fruit in New Zealand, and dug ditches in Germany."

GARY WINTZ

TRAVEL CONSULTANT

Santa Monica, California

WHAT I DO:

I'm involved in many phases of the travel business. I'm a consultant, a writer, a lecturer, a photographer, and a tour director. I've led tours to almost every country in Asia. My specialty areas are Tibet, China, Cambodia, Laos, Vietnam, and Mongolia.

As a consultant, I design tours for travel companies. Now, for example, I'm working with a number of companies setting up tours to Vietnam and Laos. Over the last few years, there's been a gradual relaxation of travel restrictions to those countries, and now tour operators want to get in. I'm advising them on how to do it, and I'll probably write and provide photographs for their brochures. I may also lead the first few tours, before training other guides to replace me.

Gary reads an article on Tibet before planning a tour there.

HOW I GOT STARTED:

When I was 12, I started saving money for a trip to Europe when I graduated from high school. Six years later, I took that trip, hitchhiking from place to place. It changed my life. I knew then that the rest of my life would be involved with travel because it opened my mind and stimulated me so much.

After I graduated from college with a degree in philosophy, I saved up enough money to do some serious traveling. I took a four-year, around-the-world trip. I taught English in Kyoto, picked fruit in New Zealand, and dug ditches in Germany.

Then in 1981, after I got married, my wife and I accepted an invitation to teach English at a university in China. While we were there, we ran cultural training seminars for professors and researchers who were

Gary describes the route that a tour group will take.

about to leave China. We taught them how to adapt to life in the United States. After that, we were invited to teach in Tibet.

A couple of years later, I met someone who was running bicycle tours to China. Because I had lived in China, traveled widely there, and could speak some Chinese, he asked me to lead a couple of the tours. I quickly gained a reputation as a skillful tour leader.

HOW I FEEL ABOUT IT:
I'm an individualist. I don't like the idea of a nine-to-five job. I like travel and sharing the world with people. For me, it's almost a spiritual vocation to open up the world to people who may not have had a chance to see it before.

There are some problems associated with being a tour director, however. If things go wrong, even if they're beyond my control, I'm the first person the tourists blame. And it's also hard work for not great pay. That's why most tour leaders work for rewards other than money.

WHAT YOU SHOULD KNOW:
To work in the travel field, you have to have an understanding and appreciation of other languages and cultures. Study international subjects — political science, anthropology, history — and study foreign languages.

I can speak some Chinese, German, and Indonesian. It's not essential to know the language of every country to which you'll be leading tours, but it helps.

If you want to be a tour leader, you must be an outgoing person, a natural leader capable of taking responsibility and making quick decisions. Anything can happen on a tour. You have to be the shepherd of your flock. And you also have to be a disciplinarian.

I once had a man along on a tour who used foul language and abused women. He was an important businessman, had paid a lot of money for the trip, and he was used to being the boss.

But I had to kick him off the tour. You should always try to find a positive way to deal with a problem, but sometimes you have no choice.

Tour leaders who are just starting out receive little or no pay. The average tour company pays its leaders about $50 a day. But they're given the trip free.

If you're a specialist, however, or if you've been with a company for a long time, you make more. And if you work full time, you earn a salary. But the money is never great. Most tour directors make up for this in other ways. For example, when you're in Nepal, you can buy a carpet for $200 that you can sell here for $600.

Gary answers some final questions at the airport.

"I've seen more heads of state than I can count."

KATHRYN WAHLIN
TOUR GUIDE
New York, New York

WHAT I DO:
I give tours at the United
Nations in both Swedish and
English. I take people into
the major council chambers
and explain to them the work
that goes on here. The UN is
involved in many areas of
international relations, and
I have to be knowledgeable
about all of them. I have to
be able to speak and answer
questions on international
conflict resolution, disarma-
ment, economic and social
development, the fight against
hunger and homelessness,
medical care for children,
and the environment — to
name just a few of the areas
in which the UN operates.
It's a very demanding job.

I arrive here at 9:15 A.M.
for the daily twenty-minute
press briefing on recent
events in the world and
recent activities at the UN.

Kathryn conducts a group of
Swedish students on a tour.

Then I go with the other
guides to the lounge, where
we wait to be called up for a
tour. That might be right
away, or it might take half
an hour. Each guide here
has to speak English and at
least one other language.
Altogether, there are eighteen
languages in which the UN
offers tours.

Every tour is different. It
might be a group of kinder-
gartners, one of university
students, or one that includes
a mixture of all ages. I give
as many as five one-hour
tours each day.

HOW I GOT STARTED:
When I was in the fifth grade,
I did a report on my Swedish
heritage, and I decided that
one day I would live in
Sweden. In college, I was
able to spend my junior year
abroad studying at a Swedish
school. After a year there,
I had learned the language
fluently.

When I finished college, I looked for a job that would allow me to use my language skills. I answered an ad in the *New York Times* for UN tour guides who spoke Swedish. A lot of Swedish tourists come to America, but there aren't all that many Americans who speak Swedish. I'm rare — most of the Swedish guides here are native Swedes. After I was interviewed and hired, I went through a very intensive training program, during which I learned how the UN is organized.

HOW I FEEL ABOUT IT:
It's gratifying when I'm able to answer questions and help people learn more about international affairs and how the UN works. Another wonderful thing is that I get to meet and work with people from all over the world. There are guides here from Israel, Japan, France, Morocco, Ghana — everywhere. I've gotten to meet an incredible group of people.

Another big advantage to working here is that I get to see some fascinating things. I've seen more heads of state and ministers than I can count. In 1990, for example, I got to see Nelson Mandela from South Africa speak, and he was very impressive. You could see his inner strength as he spoke. Later that year was the World Summit for Children, for which seventy-

Kathryn describes a new exhibit to a tour group.

Kathryn answers questions at the end of each tour.

one heads of state gathered in the same room. That was really exciting.

I was also able to see a lot of the meetings in which the UN debated the situation in the Persian Gulf, such as the sessions in which the member nations decided whether or not to allow the use of force. Unless a meeting is closed, staff members can show their ID cards to the guards and watch the proceedings. I do this when I have some free time.

WHAT YOU SHOULD KNOW: For guide work, there is no special degree required. The main requirement is fluency in a second language. That's why I'd recommend that anyone interested in this work take the opportunity to spend a year studying in a foreign country, because living in another country is the best way to become fluent in its language. You will also get to see another way of life, to see the world from a different perspective.

You could do guide work as a career, or you could use it to learn about international affairs and then move on to another job at the UN or outside. You could move into public relations or into a job focusing, say, on apartheid in South Africa or the status of women around the world.

At the UN, the pay is very good. A beginning tour guide earns $12 an hour, and full-time staff members also make very good salaries.

"Language skills are important, but you always need something else."

MOLLY BRENNAN

FABRIC IMPORTER

New York, New York

WHAT I DO:
I'm the national sales manager of an international French textile company. We sell fabrics to interior designers and architects for use in hotels, homes, and offices. I'm responsible for our showrooms in the United States, and I also relay news about the American operation back to France.

Of the eighteen showrooms I oversee, two are owned by the company, so for those two I'm in charge of the hiring and firing of staff. The other sixteen showrooms are independently owned. At these independent showrooms, I work with the showroom managers, teaching them about the fabrics and coaching them in sales techniques.

My job involves a lot of traveling and a lot of meet-

Molly examines a pattern from the new collection of fabrics.

ings. When I meet with the local showroom managers, we discuss how business is going, and sometimes we meet with important customers to display our latest collections.

I also travel to France to meet with the national sales managers for other countries. At these international sales meetings, which last from three to five days, we usually speak French, but sometimes we speak English. First the new collections are presented by the designer who owns the company. Then the sales force discusses the business strategy for the next six months, while the designer takes our questions and comments on our approach.

International sales meetings are usually very interesting. It's fascinating to get together with people from Cyprus, Lebanon, England, Italy, Japan, and

Molly looks over sample swatches with a client.

other countries, who are selling the same fabric that I'm selling. I also meet with these people on a social basis outside the conference. We speak French or English as we discuss common problems and strategies. We share ideas and talk about what it's like to work for this company in our different countries.

HOW I GOT STARTED:

I lived in France for a while, and when I returned to the United States, I wanted to find a job using my French skills. I decided to move to New York City because I knew there were bilingual jobs available there.

I signed up with a few employment agencies that place people with foreign language skills, and they got me interviews with companies that needed French-speaking employees. I interviewed with several banks and corporations, but I really wanted a job that combined business with my interest in art, so I chose this job. Here I get to be involved with interior design and still be on the business end of things.

I was hired for this job because I spoke French well, but also because my liberal arts education gave me a strong general background. After I had been working here for six months, though, I had to take a course in the technology and design of textiles. I've been here eight years now.

HOW I FEEL ABOUT IT:

I like my work very much. I really enjoy the responsibility. I started out as an assistant in a showroom, then I became a showroom manager, and now I'm the national manager.

I love traveling to the different showrooms, which are located all over the United States, as well as taking occasional trips to France. It's very stimulating, but it can also be tiring.

WHAT YOU SHOULD KNOW:

Keep up your interest in language, but let it be a secondary focus. Make sure that each stage in your education brings you knowledge in a new area. Language skills are important, but you always need something else, whether it be computer skills, a knowledge of art, business training, or whatever. There's a lot involved in any job, so get a liberal arts education and try to take as broad a range of courses as you can.

As far as the money goes, a starting salary in my field is about $18,000 a year. After several years, you can make $35,000 a year, plus commissions. And if you're working on a national scale, your commissions will be higher than in a regional territory.

Molly discusses sales with a showroom manager.

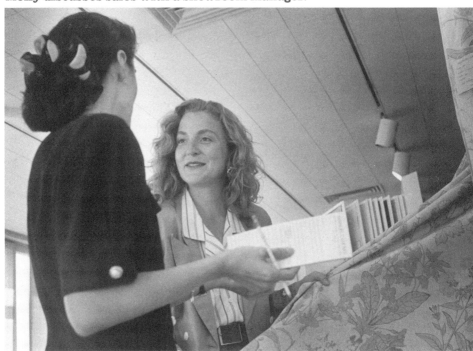

"I grew up in a bilingual community."

GERALD CURTIS

LANGUAGE TEACHER

Miami, Florida

WHAT I DO:

I teach undergraduate and graduate courses in Spanish and Portuguese at the University of Miami. This semester, I have two classes: advanced Spanish and intermediate Portuguese. But that's not a normal load. My course load was reduced this semester because I took on extra duties as the administrative assistant to the chairman of our department.

This morning, I had to deal with several problems that are part of my routine. A girl wanted to transfer her credits in Ukrainian from Harvard to Miami, but we don't teach Ukrainian here, so I had discuss her request with the dean. Then I had to ask one of our teachers to refund his travel expenses for a convention that bad weather forced him to miss.

Gerald teaches an advanced Spanish language class.

This is also the advising period for students who will be returning in the fall, so I'm busy doing that as well.

Besides the actual teaching, I also spend time preparing for class. The Portuguese conversation class is fairly easy to prepare for because the textbook suggests a variety of activities. I have to go over the lesson in advance, however, to make sure that I understand all the material and that there are no linguistic problems in the lesson. The textbook is oriented toward Brazil, and there are slang expressions in it that aren't in the dictionary. Sometimes I have to call a local Brazilian lady to find out their meaning. And sometimes even she doesn't know.

Preparation for my Spanish class is much more lengthy because it involves Spanish history and civilization, so I have to be prepared to discuss Spanish music,

23

Gerald stops to talk to one of his many students.

art, history, and culture. This entails a lot of work.

HOW I GOT STARTED:

I grew up in Arizona, in a bilingual community where people spoke both Spanish and English. I also studied Spanish in high school. Then, in college, I took time off to do missionary work for my church. I was sent to a group of Spanish-speaking people, so I got to learn the language well.

Later, in the army, I was stationed in Germany, and I studied German there. It was then that I decided I wanted to go into teaching language. I received a fellow-ship to do graduate work at the University of New Mexico, and I got my degree in Ibero-American studies. *Ibero* refers to the Iberian penin-sula of Spain and Portugal. I started teaching in 1963 at the University of Kansas, where I taught for six years. Then I came here.

HOW I FEEL ABOUT IT:

In spite of the fact that I'm often teaching the same material, there's always something new and challeng-ing. The students are always different, and there's great pleasure to be had in seeing them grow more capable, partly through my efforts.

The research environment at the university is also very stimulating, because research provides the opportunity to participate in the learning process at a higher level. What you learn from your research you can both publish and share with your students. But there is a problem here. The time you spend doing research is time taken away from your teaching. It's often hard for college teachers to find a comfortable balance.

The main problem for me is dealing with exams and grades. Just as it's fun to give good grades, it's unpleasant to give bad grades because you're recognizing a failure. And while a poor grade might be a valuable learning experience for a student, it's nevertheless negative in the short term.

WHAT YOU SHOULD KNOW: Experience abroad is invaluable to a language student or a language teacher. Spend some time, a summer or a year, in the country whose language you're studying. It's not only a great linguistic experience but a marvelous cultural experience as well. It will solidify your knowledge and change your life.

The money in college teaching depends on the department. Science departments are generally able to pay better salaries than are language departments. A newly minted Ph.D. in foreign language can expect to earn between $25,000 and $35,000 the first year.

Gerald researches each of his Portuguese lesson plans.

"There can be as many translations as there are translators."

LEE FAHNESTOCK
LITERARY TRANSLATOR

New York, New York

WHAT I DO:
I translate French books into English — mostly poetry, short stories, and novels, as opposed to technical manuals. The process typically begins when an editor contacts me. At first, I do a sample to see whether the book suits my interests and abilities and whether the editor likes my work. Then, if the sample is suitable, I sign an agreement with the editor and go to work.

For the first draft of a translation, I follow the original manuscript as closely as possible so that I can get its literal meaning right. After I've finished, the editor reads the draft and points out sections where the text might read a little awkwardly. Some passages, for example, might sound like spoken English rather

Lee looks over a manuscript that she is translating.

than written English. Then I let the book sit, so I can come back to it with a fresh eye.

Working on the second draft, I look for sections in the first draft that don't make sense or aren't good English. Also, I may have translated some things incorrectly, such as a word that has two meanings, both of which seem to fit. An example of this is the French word *grand,* which can mean either "great" or "big."

I do all this work on a computer because translation involves a lot of revisions, which are easier to make on a computer. When I've finished, I read the manuscript aloud to someone. That's a good final test.

HOW I GOT STARTED:
Both my parents spoke French, so I heard it all during my childhood. When I studied French in high

school, it came naturally to me. Later, while spending a year in Italy, I learned Italian, and now I also do some Italian translations.

I came to translation rather late in life, however. I raised a family, then I went back to school. I got a master's degree in comparative literature and attended translation workshops, where you read your translations to other students and in this way begin to get your voice. Like a writer, a translator must develop a voice, which is the way you sound on the page.

HOW I FEEL ABOUT IT:
I think of translation as writing on a tether. You're attached to the original work, but you're adapting it and bringing it into another medium. Some people think translation is making a copy. But that's not so. There are close translations, and then there are loose translations. There can be as many translations of an original work into English as there are translators.

When the translation works really well, it's exhilarating. I've had some authors say, "That's the way I would have written it if I'd written in English." On the other hand, there's always the fear that you'll turn out something that just doesn't work, or that you've misread or mistranslated something.

Lee considers the best way to translate a French phrase.

Lee works on the translation of an Italian novel.

WHAT YOU SHOULD KNOW:
People often ask me, "How well do you need to know a foreign language to be a translator?" I think it's better to ask how well you need to know your mother tongue. You always translate into your mother tongue, and the value of your translation is finally dependent on your ability with your native language.

Still, a feel for the foreign language is an important thing. You must have it in your ear. Translation workshops at universities can help you work on this sort of thing, giving you both confidence and practical knowledge. You can also go to a library and find copies of a work both in the original language and in English.

Then make your own translation of the original and compare it to the English version from the library. There will be differences in style, of course. Maybe yours will be more contemporary.

For book translations, the amount you get paid depends on the difficulty of the work and also on the status of the author and the publisher. A rough figure for an experienced translator is $70 per thousand words. This works out to about $3,000 for a short book, which might take six weeks. Occasionally, a translator gets royalties, but that's rare. I once did an updating of a translation of Victor Hugo's novel *Les Miserables*. It was 1,220 pages long, and I was able to get royalties for my revision.

"People in the audience wearing earphones hear my voice."

EVA DESROSIERS

INTERPRETER

Falls Church, Virginia

WHAT I DO:
I'm a simultaneous inter-
preter, which means that I
sit in a booth and interpret
what a speaker is saying as
soon as he or she says it.
People in the audience wearing earphones hear my voice.
I work for several different
international organizations,
including the Organization
of American States, the Pan-
American, and the State
Department. I translate
Spanish into English.

I'm also a federally
certified court interpreter.
This certification allows me
to work in federal and state
courts, interpreting the
words of defendants who are
not English speakers. At the
same time, I interpret for
the defendants everything
said by the judge and the
attorneys. If there are

Eva performs simultaneous
interpretation of a speech
at the State Department.

Spanish-speaking witnesses,
I also interpret what they say.

Another kind of interpreting I do is called consecutive
interpreting. For example,
when the President of the
United States goes on a trip,
he takes along a consecutive
interpreter. He speaks a few
sentences, and then he waits
while the interpreter trans-
lates those words into
another language.

Sometimes my work in-
volves both consecutive and
simultaneous interpreting.
I might work at a military
conference, for instance, and
during the day, I'll sit in a
booth and do simultaneous
interpreting. Then, at dinner,
there will be a guest speaker,
for whom I might do consecutive interpreting.

HOW I GOT STARTED:
I grew up speaking three
languages: English, Spanish,
and German. After I finished
school, I worked for an inter-

31

Eva adjusts the volume controls on a microphone.

national organization as a bilingual secretary. My boss wanted me to translate and interpret for him, but I was still called a secretary, because in those days women were supposed to be secretaries. Later, my husband was sent by the government to Central America, and that's where I started doing serious interpreting.

HOW I FEEL ABOUT IT:
Even when the subject is boring, this is exciting work because you always have to come up with the right word or sentence. It's challenging because your translation has to make sense both literally and in terms of the style of the speaker's language. In court, if somebody is speaking well, I try to maintain that. If, however, the person testifying does not speak his own language well, I try to convey that also. For example, if someone said in English, "I done it," I wouldn't translate that sentence into the proper Spanish. Instead, I'd match the literacy level to convey that these are not the words of a highly educated person. It's very important to project not your own personality, but the personality of the speaker.

The problem with being a freelance interpreter is that

you can't make any plans. You always have to be available because days you don't work are days you don't make money. Also, in any interpreting work, there's always the fear that you'll misinterpret something. Usually we work in teams of two, however — half an hour on, half an hour off — because your brain has to take a break. Also if you miss a word, your partner can point it out to you.

WHAT YOU SHOULD KNOW:
In order to become an interpreter, you have to be bilingual. But it's important to be bicultural as well because words are not enough sometimes. You should also understand the cultures involved.

You don't need a particular degree for this work, but you do have to pass tests given by organizations such as the State Department and United Nations. And you should also be able to work under pressure and prepare for a new assignment on very short notice. To interpret in courts, you have to pass very difficult written and oral tests. Only about 10 percent of the people who take these tests pass them.

When interpreters work, they make good money. Conference work pays $350 per day or more, and courtroom work pays $250 per day. A day is usually six hours long. And if you have to travel, your travel is paid for, too. On the other hand, a full-time senior interpreter for a large international organization makes at least $50,000 a year.

Eva reviews the text of a speech being given in Spanish.

RUSSELL G. SHUH

LINGUIST

Los Angeles, California

WHAT I DO:
I'm the chairman of the department of linguistics at the University of California at Los Angeles. Linguistics is the study of the general structure of language, the features that all languages have in common. It's divided into several areas: phonology, the study of sounds; morphology, the way words are put together; syntax, the way sentences are put together; and semantics, the things that words and sentences mean. Linguists usually specialize in one of these areas.

My specialty is African languages. There are about two thousand different languages spoken in Africa, and I've worked with about twenty of them. I collect information on these languages by working with

native speakers. Then I try to describe what each language is like, what its sounds are, how its words are put together, and how sentences are put together.

I'm also interested in how languages relate to each other, especially over time. Shakespeare's English was different from our English, and we can see the differences in books and manuscripts. But many languages have no written record, which means that the only way to explore their histories is to relate them to one another. French, Spanish, and Italian, for example, are similar because they all came from Latin. Therefore, if three African languages are similar, it may be reasonable to conclude that they all came from one common language as well. Part of my work involves trying to reconstruct what that original language might have been.

Russell analyzes the sounds of the languages he studies.

HOW I GOT STARTED:

My story is typical of a lot of people in linguistics. I've always been interested in language. I took Latin in high school and loved it. In college, I got a bachelor's degree and a master's degree in French.

But degrees in foreign languages are usually degrees in the literature of those languages. Personally, I didn't care for the study of literature. What I liked instead were the nuts and bolts of language. When I did my undergraduate work in the early 1960s, there weren't many linguistics departments. But from one of my professors I found out about the study of linguistics. And I said, "That's what I want to do." So I went to the University of California at Berkeley and studied linguistics there. After graduating, I went into the Peace Corps because I knew I would get to study an exotic language that way. I was sent to Africa, where I worked with several languages. Then I came back and got my Ph.D. at UCLA.

HOW I FEEL ABOUT IT:

I love to mess around with languages, and I'd probably do it even if I didn't get paid for it. The advancement of knowledge is what I hope to achieve. I work with languages that nobody in the outside world would know about if it weren't for my interest. Also, linguistics has many useful applications. Linguists are involved in bilingual education, for example, and in determining the best way to teach foreign languages.

The biggest problem for me is that linguistics is mostly an academic pursuit, in the ivory tower and separated from the real world. What do Iraqi refugees care that somebody at UCLA is studying African languages? Linguistics isn't a tangible thing. People can't easily understand its benefits, and sometimes I wonder about doing work that doesn't have an immediate practical use.

WHAT YOU SHOULD KNOW: Linguistics is a formal field, so it's useful to have the kind of mind that enjoys mathematical or scientific reasoning. Also, you should be fascinated by language. Study languages in high school; learn as many foreign languages as you can. Then apply to a university with a linguistics department.

Financially, academic jobs aren't the highest paying in the world, but they aren't bad either. As a university professor, you'll never starve, and you'll have regular promotions. Universities also tend to be nice places to work. For instance, I have good medical benefits, access to a wonderful library, and I work with interesting people. The starting salary for an assistant professor with a Ph.D. is in the low thirties.

Russell consults with two of his graduate students.

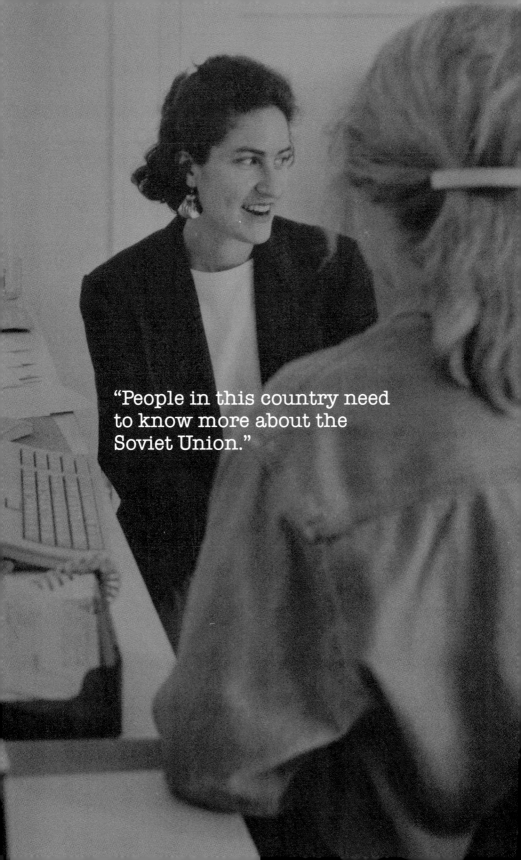

"People in this country need to know more about the Soviet Union."

JAIMIE SANFORD

INTERNATIONAL BUSINESS FACILITATOR

San Francisco, California

WHAT I DO:
I work for a nonprofit organization that focuses on U.S.-Soviet relations. We run a program for aspiring Soviet entrepreneurs, which offers them management training in the United States through one-month internships with American companies.

There are a great many Soviets who have expertise in a given field but don't know how to go out and start a business. They just don't have the organizational and management skills, so we help them learn those skills. We might help a lawyer, for example, who wants to go into private practice. Or we might help someone who wants to open a hotel, a bakery, or a restaurant.

I do everything from organizing our computer data base to writing materials

Jaimie arranges internships for Soviet businessmen.

for use in the program. I write application forms and orientation materials, which tell the Soviets what to expect when they arrive in the United States as well as what to expect from American businessmen. I also create orientation materials for the Americans, to give them an idea of what to expect from the Soviets whom they have agreed to take into their homes and businesses. My own background in Russian comes in very handy here and so does my knowledge of the Soviet Union.

A particular responsibility of mine is marketing the program to both Soviets and Americans. I oversee all the applications as well as the screening process by which we pick qualified Soviets to come here. I also educate the managers of American businesses about the program.

HOW I GOT STARTED:

I wanted to study a difficult language in college, one that would challenge me and expose me to people I would never meet otherwise. Also, I was quite involved with anti-nuclear protests at the time, trying to figure out why the Soviets and the Americans were building so many nuclear weapons. I felt very strongly that people in this country needed to know more about the Soviet Union. So I decided that to learn more about the Soviets, I would become a Russian studies major.

I was lucky because Gorbachev came to power not long after I graduated, and suddenly there were opportunities to work with the Soviets in ways that hadn't been possible for decades. Travel and communication opened up, and within a couple of years, there was mass support in the United States for contact with the Soviets on many different levels.

After graduating from college, I worked for a tour operator who specialized in the Soviet Union, and I did troubleshooting and advance work both in Washington, D.C., and over there. Through this experience, I learned what Americans need to know when they go to the Soviet Union, what they find interesting, and what Soviets think about the United States.

HOW I FEEL ABOUT IT:

My work is exciting because I have the opportunity to develop a whole new program. A lot of my tasks are administrative, though, which can be boring. And I don't like having to adhere to the conservative dress code and nine-to-five routine of the business world. But I always keep in mind that I'm helping change people's lives.

Jaimie screens all of the applicants to the program.

Jaimie teaches American sponsors about the Soviet Union.

Right now, in the Soviet Union, people are realizing that they have to move toward a free-market economy. And they're looking to the West for economic models. That's why it's important for the United States to show them the best and the worst of what we have, so that they can develop their own system without making the same mistakes we have.

WHAT YOU SHOULD KNOW:
I think studying a difficult language is one of the greatest things you can do, because you can easily see your progress. You learn the alphabet, then you learn phrases, and finally you can look back at what used to be meaningless scribble, and it makes sense. It's exciting — like breaking a code.

The money isn't great because this is the nonprofit world. But the skills you develop are the same ones needed by for-profit companies, and they can always be transferred. I'm not making great money, but I enjoy what I do, and I can move on later to a job that will pay more. A starting salary in this field is about $21,000 a year.

"Many foreigners think that America is only Disneyland and 'Dallas.'"

LIBBY CRYER

FOREIGN EXCHANGE DIRECTOR

Greenwich, Connecticut

WHAT I DO:

I'm the director of Academic Year in America, a program run by the American Institute for Foreign Study (AIFS) that enables 15- to 18-year olds from foreign countries to spend a semester or an entire school year in the United States. The students live with American families and attend local high schools.

The program is implemented by a network of volunteers, but the volunteers are managed by a national staff, which I direct. My job involves hiring and supervising about twenty-five paid employees, coordinating with other groups involved with foreign exchange, and dealing on a daily basis with our seventy-five overseas partner organizations in twenty-five countries.

Libby finds host families for foreign students who come to study in the United States.

A large part of my job involves dealing with cross-cultural differences, especially in trying to explain our program to foreigners. Even the most fluent English-speaking people have the most incredible misconceptions about America. Many think that America is only Disneyland and "Dallas." They think American Moms stay home, American Dads are the breadwinners, and every American house has a swimming pool and two or three cars in the garage. But we place kids in all kinds of homes, often with single-parent families, often with blue-collar families. And I have to spend a lot of time explaining this to our overseas partners.

Because I speak French and a little Italian, I sometimes counsel French or Italian students who are having difficulty adjusting to American life. At first, I try

43

Libby reviews a program brochure with colleagues.

speaking with them in English, because they're here to learn the language. But if they're really upset, I switch to their native language to make them feel more comfortable. I also speak French daily with French exchange groups, and I'll occasionally talk to a parent in France.

I also travel a fair amount so that I can make personal contact with potential partner organizations. Often the leaders of these organizations speak English, but usually the people in the office do not. It really makes a difference to be able to go into their offices and speak their language with them.

HOW I GOT STARTED:
I was an exchange student once myself. I spent my junior year in college at the University of Geneva, which is how I became interested in foreign exchange.

I graduated with a degree in international banking, but I found banking deadly dull. I left to raise my children, and then when I reentered the job market, I took a job with AIFS.

Our program, begun in 1978, is booming now because there is incredible international interest in learning English. Kids from other countries want to come here so they can learn English while they're young. Also, the dollar is weak abroad, so people can afford to come.

HOW I FEEL ABOUT IT:
This is a program that really changes lives, both of the

kids who come here and of the American host families. Friendships are formed that last a lifetime. Perceptions are changed in the most profound ways. You can go to a travel agency and buy a vacation, but you can't buy the experience of being a child in an American family.

The kids say the most amazing things about their experiences — how wrong their expectations were, what they think it means to be an American. In this way, we're a grass-roots peace movement. The host families discover that kids are kids no matter where they come from. They all wear jeans and leave their shoes in the middle of the floor. And the kids discover that Americans have emotions and dreams and problems just like everybody else in the world.

WHAT YOU SHOULD KNOW:
If you think you want to do international work, you should travel or work abroad. My own daughters have discovered opportunities abroad, such as doing au pair work, which is working for a family in exchange for room, board, and a small salary. Having experience abroad sensitizes people.

In terms of money, we like to think we're on a par with others in the field. But because we're a non-profit organization, we're not always competitive with the for-profit world. The starting salary for one of our regional directors is $29,500 a year, and an entry-level position pays about $23,000 a year. But you also get the satisfaction of changing lives and making the world a better place.

Libby posts forms filled out by students seeking hosts.

"Yesterday, we had ten Polish publishers here."

TIMOTHY ROGUS

FOREIGN LANGUAGE EDITOR

Lincolnwood, Illinois

WHAT I DO:
I work with foreign language textbooks, evaluating proposals for new books, issuing contracts, and overseeing the manuscripts as they pass through production. I specialize in the European languages. My strongest languages are French, Spanish, and Portuguese, but I also know enough Italian and German to read and evaluate proposals.

I work for a publisher of textbooks that's middle-sized but with a rather commanding position in the field. We get some of our books from writers who send us proposals, but others we buy overseas and then adapt to the American market. Besides textbooks, we also publish readers, workbooks, cassettes, and other classroom materials.

Timothy works on the contract for a French language text.

My job used to involve a lot of day-to-day work on manuscripts, but I do less of that these days. Instead, we use a lot of freelancers to edit manuscripts, after which I evaluate their work page by page. I may change things they've done or add new things.

I also deal with our graphics department on design and illustration, especially when it comes to hunting for photos to illustrate a book. And I'm constantly talking to authors on the phone. Often they have questions about what tack to take while they're writing the book.

The books I oversee range in level from alphabet books in French and Spanish to advanced readers for college students with articles from French and German magazines. Many of these books contain activities and review questions, which are often

Timothy evaluates a proposal for a revision of a textbook.

developed in collaboration with the authors.

HOW I GOT STARTED:

I started learning French in the sixth grade and continued my studies throughout high school and college. I also took Latin in high school, and along the way I studied other languages, too. In college, I decided to major in French, and I even went on to get a master's degree in the language.

I taught French in college for a while, but I didn't par-ticularly like teaching, even though I still loved foreign languages. Fortunately, I found in publishing a way to pursue my love of languages while at the same time utilizing my teaching background. I'm still teaching, just not in the classroom.

HOW I FEEL ABOUT IT:

I like my job. The variety is incredible — sometimes over-whelming — but that's what keeps it interesting. I get to attend foreign language con-ferences around the country, I travel to New York at least once every year, and I've gone to the Frankfurt Book Fair in Germany twice. I really like all of the travel. But I also enjoy sitting down with a pencil and improving a manuscript.

Another thing I enjoy is dealing with the authors and publishers from all over the world. Yesterday, for instance, we had ten Polish publishers here, and I talked on the phone with a Russian pub-lisher today. I've always liked dealing with people from other countries.

I don't especially care for the administrative work, however. I have to fill out purchase orders and invoices and write editorial previews for catalog copywriters. These sorts of tasks I find tedious, but it's all part of the job.

48

WHAT YOU SHOULD KNOW:
You have to have a solid background in one or two foreign languages to work in this field, but you also need a solid background in English. You have to know how to write well in English because you may have to write introductions, back cover copy, or what have you. And you have to be able to judge other people's writing.

To get a job in textbooks, it helps to have some teaching experience. Often textbook companies look for people who've done some teaching because to know what works in a classroom, it helps to have spent some time there yourself.

In terms of salary, you start off relatively low in publishing, but with time and experience, the pay gets better. The company I work for pays fairly well, about $18,000 a year to start. I've been here for seven years, and I earn $38,000. It's not the salary of an engineer or a physicist, but I enjoy what I'm doing.

Timothy looks at some illustrations with a designer.

49

"Language is more than just words. It's feelings and actions."

MOLLY McGINN

CROSS-CULTURAL TRAINER

Chicago, Illinois

WHAT I DO:
I develop language-learning materials for a major airline, including audio tapes, video tapes, and books. I also supervise the cross-cultural training program and train the trainers. Sometimes I even teach classes.

I go to Asia quite often and conduct training programs in Singapore, Hong Kong, and Japan. Over there, it's my job to give our staff a broader understanding of the cultural background of our customers so that they can better understand and anticipate their needs.

For example, Japanese passengers on an airplane will behave in one particular way, while Indian passengers will behave in a quite different way. We try to make the flight attendants aware of these differences, what

Molly often travels overseas to conduct training programs.

they mean, and how to respond to them.

There are also differences in the way people communicate. Japanese, for example, don't say *no* with words. They wave one hand back and forth in front of their faces. Flight attendants need to know that this means "No, thank you." Asking customers what they mean isn't enough. If a customer doesn't understand English, he or she might not understand what the flight attendant is asking.

HOW I GOT STARTED:
I got a bachelor's degree in psychology because I've always been interested in how the mind works. Then I continued my studies in Kyoto, Japan, where I lived for three years in a Zen monastery to get a sense of the Eastern approach to psychology. While I was there, I taught English and worked for a Japanese com-

51

Molly talks to two flight attendants after a seminar.

pany. After that, I returned to America and worked as a language trainer before I got my first job with an airline as a language-qualified flight attendant.

Later, after studying Chinese, I took a leave of absence and worked for the Chinese government in China teaching English. I also worked in Tibet for a while. Knowing Chinese and Japanese helped me to get a doctoral program in psychological anthropology, which is the study of how culture affects the way people think. After that, I was hired by this airline.

HOW I FEEL ABOUT IT:
The flight attendants in my classroom are going to be on 747s headed for Singapore or Paris the next day, so they really need the information I'm giving them. And I get satisfaction because they genuinely want to know how they can do a better job dealing with people from all over the world.

Learning a new language is a key that can open doors to other worlds. It's very rewarding to have basic interactions with people who don't speak English, to be able to say, "Would you care for more coffee?" in Japanese and have people understand you and respond. It's as though you've broken a secret code, and it gives you a sense of connectedness to other people, no matter where they're from.

The biggest drawback is being away from home so much. Last year, I was at home for only about fifty days. I was doing fieldwork in Japan part of the time, and I'm often traveling to do training in other countries.

WHAT YOU SHOULD KNOW:
Whatever country you're interested in, get a backpack and a round-trip ticket and go there. Language is more than just words. It's feelings and actions, and it should be learned in context, in the country where it's spoken. It's fun to travel with a friend, but you'll meet more people if you go alone. Don't be afraid of the unknown. People will always help you along the way.

Becoming a flight attendant is a particularly good doorway into the airline business, or into international business in general. Starting pay for flight attendants is low — about $1,200 a month plus benefits. But after a few years, you can be making a lot more. And you can move up from there, either in the airline business or working for another corporation.

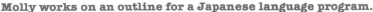

Molly works on an outline for a Japanese language program.

"Working only in America would be like going deaf in one ear."

MARIA CAMPBELL

INTERNATIONAL LITERARY SCOUT

New York, New York

WHAT I DO:

My job is to scout American books, fiction and nonfiction, for foreign publishers. I work with publishing houses in ten countries, keeping them informed about potential American candidates for their lists. Recent books that I've placed with foreign publishers include *Presumed Innocent* by Scott Turow — which was bought by publishers in Holland, Germany, Sweden, and Italy — and *The Joy Luck Club* by Amy Tan, which was purchased by an Italian publisher.

A typical day for me begins with a stack of faxes relating to a number of pressing concerns, such as the foreign reaction to news of an important book. The fax machine has sped up the entire scouting field. Its immediacy gives an extra urgency to the work.

Maria considers a new American novel for one of her clients.

Three times a week on average, I'll have business lunches with editors, authors' agents, journalists, and subsidiary rights people to keep up with what's coming out. Subsidiary rights to a book include the right to publish excerpts in a magazine and the right to make a movie out of the work, as well as the right to publish a foreign edition. After lunch, I might meet with an editor, an agent, or a foreign publisher. Or I might go to an author's reading.

Another part of my job involves assisting foreign publishers when they come to New York. I schedule meetings for them with American publishers and give them general support and guidance. I might even suggest places for them to take their children while they're here.

The role of the international literary scout has gotten a lot more attention

Maria and her staff keep up with the latest books.

in the last five years as American publishers get bought out and the domestic market diminishes. There are probably twenty foreign scouts in this country, all in New York.

HOW I GOT STARTED:

My family is Italian, and I grew up speaking Italian, which pointed me in an international direction. I also studied French in school.

I started out working for an Italian publisher, at first doing translations and then scouting some American books. Over the course of the next few years, other international publishers became aware of what I was doing, and I began scouting for them as well. In time, I turned scouting into a regular job.

HOW I FEEL ABOUT IT:

I love working with books internationally, because I'm constantly learning. When I started working for the Scandinavian market, for example, I knew nothing about it. But my clients educated me, and they continue to do so. I'm often invited to visit the countries for which I scout, where I'm introduced to writers, brought into bookstores, and shown what people there like to read. Working only in the American context would be like going deaf in one ear.

As the scout and not the foreign publisher, however, you're somewhat on the outside. It's not your decision whether or not to acquire the rights to a book. And sometimes you want to see a favorite book through editing, marketing, and publicity — but you aren't involved with that.

WHAT YOU SHOULD KNOW:
If you want to get involved in international publishing, the first thing to do is learn a foreign language well. Also, travel as much as you can. Study abroad. Work abroad. That's the most useful background you can have.

In this country, it's hard to find people who have another language and speak it well. But in Europe, everybody speaks two or three languages. America needs to educate its young people in foreign languages so that they can be part of the international order.

To learn how publishing works, an internship at a publishing house or a literary agency can be very useful, because it gives you a feel for the business. When I'm hiring for an entry-level position, I look for someone with a college degree who can write reports and speak another language. The ability to write well is key in scouting, because you have to communicate the essence of a book in your reports.

Financially, there is a greater potential in agenting and publishing than in scouting. The standard entry-level salary in publishing is about $15,000 to $20,000 a year. It's very difficult to characterize the salary for a scout because it varies so widely. For instance, you can scout for only one company and earn just $12,000 a year, or you could scout for ten and earn $120,000 a year. In general, however, scouting is not a field to pursue if you're looking to make a six-figure salary.

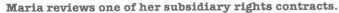
Maria reviews one of her subsidiary rights contracts.

"When I finish a case and say 'Welcome to America,' it's a very special moment."

J. BRUCE WEINMAN

IMMIGRATION LAWYER

New York, New York

WHAT I DO:
When people from other countries come to the United States and decide they want to stay here, they need to obtain visas or citizenship papers. Often, people need help with this process, and that's what I provide.

Sometimes foreigners who come here on student or tourist visas want to stay in this country longer, say, to work here for a few years. I help them get work visas. Other times people want to stay here permanently and become citizens. I help these people get temporary visas and then assist them in their applications for citizenship. About 80 percent of my practice is immigration work.

The first thing I do is interview the visa applicant to find out what his or her options are. Most people

Bruce questions the relative of a prospective immigrant.

know how they want to end up, but they don't understand their options under current law, and that's what I tell them. I sit down and make a list of the documents they need — fingerprints, photos, and so on.

My clients come from all over the world. Half of my business is with the Korean community, but I also deal with Hispanics, Indians, Africans, Chinese — you name it. People on my staff speak both Korean and Spanish, and I pay interpreters on an hourly basis when I have a client who speaks a language we don't speak here. I speak Spanish myself. I suppose it would be possible to function in this job without any foreign language, but it would be very difficult.

HOW I GOT STARTED:
At first I had a general practice, but then started to

Bruce helps a client fill out the forms needed for a visa.

lean toward immigration work as people came to me with immigration problems. My first immigration case was for my wife, who came to this country from South America. At the time I knew very little about immigration work. But in the process of doing the work for her, I learned a lot about it.

HOW I FEEL ABOUT IT:

I enjoy this work quite a bit. I like working for myself, and I like being able to help people. It's hard to put into words, but when I finish a successful case and say, "Welcome to America," it's a very special moment. And the more difficult the case, the more satisfaction I receive from a successful outcome.

WHAT YOU SHOULD KNOW:

Immigration law doesn't involve much courtroom work — it's mostly administrative — so if you're looking for that type of practice, you should probably be in criminal law. Also, if you want the big money, you should go into corporate law. But if you're interested in helping people get into this country, and if you have ties to an ethnic community, perhaps immigration law is for you. And getting a law degree is a good idea even if you don't want to practice law, because it opens up a lot of doors in other fields.

At a Wall Street firm, the starting salary is about $75,000 a year. But the Wall Street lawyers represent a tiny percentage of the total number of lawyers. In most other areas of the law, including immigration, you're lucky to get $30,000 a year to start. You can do really well financially in immigration, but it's not at that top level.

Related Careers

Here are more foreign language careers you may want to explore:

CONCIERGE

Concierges work at the front desks of hotels, supervising arrivals and departures and providing information to guests about where to eat and how to get around.

CULTURAL ANTHROPOLOGIST

Cultural anthropologists study the origins and behavior of different cultures, frequently living among native peoples and participating in their daily activities and rituals.

CURRENCY TRADER

Currency traders exchange foreign currencies on the international market.

CUSTOMS OFFICIAL

Customs officials work at border stations, checking passports and visas and looking out for goods being brought into the country illegally.

DIPLOMAT

Diplomats represent governments in foreign countries.

FOREIGN CORRESPONDENT

Foreign correspondents report news stories from the countries in which they are stationed.

FOREIGN LANGUAGE JOURNALIST

Foreign language journalists write articles for foreign language newspapers, journals, and magazines.

FOREIGN POLICY ANALYST

Foreign policy analysts make political predictions about events in foreign countries so that the government can devise effective international policies.

IMMIGRATION OFFICER

Immigration officers process applications for citizenship and make sure that foreigners living in this country have the proper documentation.

INTERNATIONAL BUSINESS REPRESENTATIVE

International business representatives work for foreign-based companies in their overseas offices.

INTERNATIONAL FLIGHT ATTENDANT

International flight attendants see to the safety and comfort of passengers aboard an airplane.

TRAVEL WRITER

Travel writers write articles for travel books and magazines that provide information about tourist attractions, accomodations, and restaurants.

Further Information
About Foreign Language Careers

Organizations

Contact these organizations for information
about the following careers:

FABRIC IMPORTER
American Association of Importers and Exporters
11 West 42nd Street, New York, NY 10036

IMMIGRATION LAWYER
American Bar Association
1800 M Street, N.W., Suite 200, Washington, DC 20036

FOREIGN LANGUAGE TEACHER
American Council on Teaching of Foreign Languages
P.O. Box 1077, Yonkers, NY 10701

INTERNATIONAL LAWYER
American Foreign Law Association
330 West 42nd Street, New York, NY 10036

FOREIGN EXCHANGE DIRECTOR
American Institute for Foreign Study
140 Greenwich Avenue, Greenwich, CT 06830

LITERARY TRANSLATOR
American Literary Translators Association
University of Texas – Dallas, Box 830688, Richardson, TX 75083

INTERPRETER
American Society of Interpreters
P.O. Box 9603, Washington, DC 20016

TRAVEL CONSULTANT
American Society of Travel Agents
1101 King Street, Suite 200, Alexandria, VA 22314

INTERNATIONAL LITERARY SCOUT
Association of American Publishers
220 East 23rd Street, New York, NY 10010

FABRIC IMPORTER
Association of Foreign Trade Representatives
P.O. Box 300, New York, NY 10024

LINGUIST
Society of American Linguists
P.O. Box 7765, Washington, DC 20044

TRAVEL CONSULTANT
United States Tour Operators Association
211 East 51st Street, Suite 12B, New York, NY 10022

Books

A BASIC GUIDE TO EXPORTING
Washington, D.C.: U.S. Department of Commerce, 1986.

CAREERS ENCYCLOPEDIA
Homewood, Ill.: Dow-Jones Irwin, 1980.

CAREERS FOR PEOPLE WHO LOVE TO TRAVEL
By Lois Dailey. New York: Prentice Hall, 1986.

CAREERS IN FOREIGN LANGUAGES: A HANDBOOK
By June Lowry Sheriff. New York: Regents Publishing, 1975.

CAREERS IN INTERNATIONAL AFFAIRS
Edited by Gerard F. Sheehan. Washington, D.C.: Georgetown University, 1982.

CAREERS IN LAW
By Charles J. Rose. New York: Julian Messner, 1983.

CAREERS IN THE HOTEL AND TOURISM SECTOR
Geneva: International Labour Office, 1976.

OCCUPATIONAL OUTLOOK HANDBOOK
Washington, D.C.: U.S. Department of Labor, 1990.

THE SCIENCE OF LINGUISTICS AND THE ART OF TRANSLATION
By Joseph L. Malone. Albany, N.Y.: State University of New York Press, 1988.

THE TEENAGER'S GUIDE TO STUDY, TRAVEL, AND ADVENTURE ABROAD
New York: St. Martin's Press, 1988.

THE TOURIST BUSINESS
By Donald E. Lundberg. New York: Van Nostrand Reinhold, 1990.

THE TRANSLATOR'S HANDBOOK
By Frederick Fuller. University Park, Pa.: Pennsylvania State University Press, 1984.

VGM CAREERS ENCYCLOPEDIA
Lincolnwood, Ill.: VGM Career Horizons, 1988.

WORK, STUDY, TRAVEL ABROAD
New York: St. Martin's Press, 1984.

Glossary Index